Humanitarian Engineering

Humanitarian Engineering

Carl Mitcham and David Muñoz

www.morganclaypool.com

ISBN: 9781608451517 paperback
ISBN: 9781608451524 ebook

DOI 10.2200/S00248ED1V01Y201006ETS012

A Publication in the Morgan & Claypool Publishers series
SYNTHESIS LECTURES ON ENGINEERS, TECHNOLOGY, AND SOCIETY

Lecture #13
Series Editor: Caroline Baillie, *University of Western Australia*
Series ISSN
Synthesis Lectures on Engineers, Technology, and Society
Print 1933-3633 Electronic 1933-3641

Synthesis Lectures on Engineers, Technology, and Society

Editor
Caroline Baillie, ***University of Western Australia***

The mission of this lecture series is to foster an understanding for engineers and scientists on the inclusive nature of their profession. The creation and proliferation of technologies needs to be inclusive as it has effects on all of humankind, regardless of national boundaries, socio-economic status, gender, race and ethnicity, or creed. The lectures will combine expertise in sociology, political economics, philosophy of science, history, engineering, engineering education, participatory research, development studies, sustainability, psychotherapy, policy studies, and epistemology. The lectures will be relevant to all engineers practicing in all parts of the world. Although written for practicing engineers and human resource trainers, it is expected that engineering, science and social science faculty in universities will find these publications an invaluable resource for students in the classroom and for further research. The goal of the series is to provide a platform for the publication of important and sometimes controversial lectures which will encourage discussion, reflection and further understanding.
The series editor will invite authors and encourage experts to recommend authors to write on a wide array of topics, focusing on the cause and effect relationships between engineers and technology, technologies and society and of society on technology and engineers. Topics will include, but are not limited to the following general areas; History of Engineering, Politics and the Engineer, Economics , Social Issues and Ethics, Women in Engineering, Creativity and Innovation, Knowledge Networks, Styles of Organization, Environmental Issues, Appropriate Technology

Humanitarian Engineering
Carl Mitcham and David Muñoz
2010

Engineering and Sustainable Community Development
Juan Lucena, Jen Schneider, and Jon A. Leydens
2010

Needs and Feasibility: A Guide for Engineers in Community Projects — The Case of Waste for Life
Caroline Baillie, Eric Feinblatt, Thimothy Thamae, and Emily Berrington
2010

Engineering and Society: Working Towards Social Justice, Part I: Engineering and Society
Caroline Baillie and George Catalano
2009

Engineering and Society: Working Towards Social Justice, Part II: Decisions in the 21st Century
George Catalano and Caroline Baillie
2009

Engineering and Society: Working Towards Social Justice, Part III: Windows on Society
Caroline Baillie and George Catalano
2009

Engineering: Women and Leadership
Corri Zoli, Shobha Bhatia, Valerie Davidson, and Kelly Rusch
2008

Bridging the Gap Between Engineering and the Global World: A Case Study of the Coconut (Coir) Fiber Industry in Kerala, India
Shobha K. Bhatia and Jennifer L. Smith
2008

Engineering and Social Justice
Donna Riley
2008

Engineering, Poverty, and the Earth
George D. Catalano
2007

Engineers within a Local and Global Society
Caroline Baillie
2006

Globalization, Engineering, and Creativity
John Reader
2006

Engineering Ethics: Peace, Justice, and the Earth
George D. Catalano
2006

Humanitarian Engineering

Carl Mitcham and David Muñoz
Colorado School of Mines

SYNTHESIS LECTURES ON ENGINEERS, TECHNOLOGY, AND SOCIETY #13

ABSTRACT

Humanitarian Engineering reviews the development of engineering as a distinct profession and of the humanitarian movement as a special socio-political practice. Having noted that the two developments were situated in the same geographical and historical space — that is, in Europe and North America beginning in the 1700s — the book argues for a mutual influence and synthesis that has previously been lacking. In this spirit, the first of two central chapters describes humanitarian engineering as the *artful drawing on science to direct the resources of nature with active compassion to meet the basic needs of all — especially the powerless, poor, or otherwise marginalized.* A second central chapter then considers strategies for education in humanitarian engineering so conceived. Two final chapters consider challenges and implications.

KEYWORDS

engineering, professional engineering, humanitarianism, humanitarian engineering, engineering education, humanitarian engineering education, engineering ethics, sustainability

Contents

Preface

This "lecture" — more accurately, small book — is about something called humanitarian engineering. It is written by an interdisciplinary scholar of science, technology, and society studies (Mitcham) and an engineer (Muñoz), both of whom are dedicated to this somewhat controversial activity.

We argue for the possibility of humanitarian engineering by considering the character of engineering in its originating historical context (Chapter 1), humanitarianism as a new historical context (Chapter 2), and connections between the two — that is, the role engineering can play in the new context in the form of the ideal of humanitarian engineering (Chapter 3). Having presented the ideal of humanitarian engineering, we then consider some key elements in humanitarian engineering education (Chapter 4) and its challenges (Chapter 5). A brief conclusion reviews the argument and places it in a broader, historico-philosophical perspective (Chapter 6).

* * *

Our basic concern has to do with expanding opportunities in engineering — opportunities that have implications for engineering education and practice. However, it must be admitted from the outset that what is written here only scratches the surface. The potentialities with which we deal are more complex than can be covered in a short text. A brief presentation may nevertheless stimulate interest and reflection.

Like engineering and humanitarianism themselves, our argument has multiple roots. Most proximately, it can be traced back to an idea by our now emerita colleague, Prof. Joan Gosink. In response to an invitation from the William and Flora Hewlett Foundation in 2002, she led the drafting of a proposal from the Colorado School of Mines (CSM) to develop an interdisciplinary undergraduate program in humanitarian engineering, with — as is necessarily the case with such dreams — only a general idea of what this might be. Our dream grew out of the assessment of a problem that was stated as follows:

> The 21st century has brought Americans a new awareness of anguish and discontent in lower income countries, and an emerging recognition of the need for U.S. participation to ameliorate this suffering. Attendant with these issues is the demand for enhanced security, safety, and equity. At the same time, engineering graduates shy away from political life, community service, and international work in the non-profit sector. Engineering talent is not making sufficient contributions toward the solutions of major human needs, particularly the need to sustain both human systems and natural systems within an ethical framework which recognizes the disproportionate impact of engineering and applied

> science in contemporary society. Further, when it comes to engineering's relevance to the world's most challenging problems, the public attitude toward engineering is not very encouraging. Leaders in engineering education and the profession have argued that the perception that engineering is irrelevant to humanity's present and future needs may be a key reason for the steady decline of engineering enrollment over the last decade, as well as the persistent under-representation of women and minorities in engineering. Engineering students are too often misperceived to be more concerned with their personal vocational interests and material goals and uncaring about society at large, particularly the plight of the developing world. [Supporting references dropped.]

In response to this assessment, our proposal, "Serving Humanity: Engineers Improving the World Through Regional, National, and International Community Service," envisioned developing

> a community service component for the . . . engineering curriculum that [would] teach engineering students how to bring technical knowledge and skill to bear on the real-world problems of the less materially advantaged in order to promote development of the common good. To accomplish this goal we [wanted to] modify existing courses and introduce new engineering courses that convey relevant knowledge and training for service missions. [The new curriculum would] consist of both technical and non-technical courses to develop the skills, expertise, understanding, and attitudes that support pro-active community service.

In 2003 the William and Flora Hewlett Foundation generously selected the CSM proposal for funding, and collaborative efforts to realize the dream were undertaken by faculty from two interdisciplinary academic units, the Division of Engineering and the Division of Liberal Arts and International Studies. By 2005 this led to creation of a new Humanitarian Engineering (HE) minor with a small group of teachers, scholars, and students who have continued ever since to pursue the ideal of creating a "cadre of engineers, sensitive to social contexts, committed and qualified to serve humanity." At the same time, the activity at CSM was not an isolated effort. Similar programs have emerged at other institutions, some of which are mentioned in our discussion of humanitarian engineering education.

* * *

From the beginning, in a situation that was also not unique to CSM, there were some difficulties in communicating both between engineering faculty involved and those not involved, and among the engineering, social science, and humanities scholars involved. For example, some faculty not involved objected to the very term "humanitarian engineering." To them it implied that other types of engineering were not for human benefit. All engineering, they maintained, was humanitarian in orientation. Indeed, even some faculty involved in the program as it has developed tend to be sympathetic with this view and, with regard to a capstone humanitarian engineering project, think almost any type of engineering project — especially, if it is for groups disadvantaged in some manner — can be classified as humanitarian. With regard to communication within the team among different disciplinary traditions, there were tensions between those who wanted to go slow, in order to incorporate knowledge and perspectives from the social sciences and the humanities, and those who wanted to get on with problem solving. To adopt a contrast of extremes: Engineering faculty sometimes perceived non-engineering faculty as "obstructionist nit-pickers"; social science and humanities faculty sometimes thought of their counterparts as "insensitive engineers." We have, however, been able to avoid letting differences devolve into excessive contention and used the natural tensions as stimulus to creative interaction.

After considerable but creative debate, we developed the following working description for the CSM humanitarian engineering program:

Humanitarian: to promote present and future well-being for the direct benefit of under-served populations.

Engineering: design under physical, political, cultural, ethical, legal, environmental, and economic constraints.

Humanitarian Engineering: design under constraints to directly improve the well-being of under-served populations.

Using this three-part descriptor, the HE team developed a minor that included three courses in Liberal Arts and International Studies (one of which is a required introduction to ethics) and a series of technical courses culminating in a two-semester capstone senior design project. One extended and successful project has involved working with a small village in Honduras to design and construct a water system, a project which was one of the 21 UNESCO/Daimler sponsored 2004/2005 Mondialogo Engineering Awards for "sustainable development and poverty reduction worldwide." The program has also been useful to help attract and retain women in the CSM engineering program.

Indeed, a supplemental goal of our humanitarian engineering educational effort was to attract students who might otherwise not see engineering as a viable career path — persons with strong aptitude in science and mathematics who also enjoy working with people and value serving others in direct ways. Such students, when considering educational options, might not realize how humanitarian work can be linked to engineering. Though difficult to quantify, our sense is that humanitarian engineering does attract such persons and that students become especially motivated by working on projects deemed altruistic or oriented toward something other than more common forms of self interest.

* * *

In an effort to address some of the paradoxes of humanitarianism — and to avoid subjecting our students to serious threats to their own safety and welfare — we have made an effort to focus not just on intervention in humanitarian crisis situations but on "preventative humanitarian action" and "humanitarian community development." Indeed, since 2005 a second interdisciplinary faculty team under the leadership of Juan Lucena (with a grant from the U.S. National Science Foundation) has worked to develop graduate courses in humanitarian engineering ethics that focus much of their attention of the relationship between engineering and sustainable community development. Another volume in the Synthesis Lectures on Engineers, Technology, and Society by our colleagues Juan Lucena, Jennifer Schneider, and Jon Leydens, *Engineering and Sustainable Community Development*, presents results from this work and thus complements the present text.

Further complementing studies are to be found in other Synthesis Lectures on Engineers, Technology, and Society. Worthy of special note are George Catalano's *Engineering, Poverty, and the Earth* (2007); and Donna Riley's *Engineering and Social Justice* (2008); along with three volumes co-authored by Caroline Baillie and George Catalano on *Engineering and Society: Working Towards Social Justice* (2009).

What follows, then, is an effort to share some small portion of our learning — a portion we hope will be of interest and benefit to others — relevant to humanitarian engineering. In this respect it is worth noting that humanitarian engineering education can have benefits well beyond immediate contexts of humanitarian crisis and need. In a progressively globalized world, the successful pursuit of industrial activity and corporate enterprise will require increased sensitivity to societal and cultural issues — precisely the kind of sensitivity that should be an inherent by-product of any humanitarian engineering teaching and learning. In the field of government service as well, the development of skills associated with humanitarian engineering can be particularly beneficial. For those students who seek to practice humanitarian engineering directly, it can be projected that numerous non-governmental organizations or NGOs will increasingly depend on the abilities of students who have contributed to and graduated from such programs. The context of engineering in the future will almost certainly include humanitarianism.

* * *

This volume significantly expands an argument initially presented in Carl Mitcham and David Muñoz, "The Humanitarian Context," in Steen Hyldgaard Christensen, Bernard Delahousse, and Martin Meganck, eds., *Engineering in Context* (Copenhagen, Denmark: Academica, 2009), pp. 183–195, and in a number of oral presentations made in various venues, 2004-2009. It also draws modestly on Mitcham, Lucena, and Suzanne Moon, "Humanitarian Science and Technology," in Mitcham, ed., *Encyclopedia of Science, Technology, and Ethics* (Detroit: Macmillan Reference, 2005), vol. 2, pp. 947–950.

Carl Mitcham and David Muñoz
June 2010

Acknowledgments

Thanks to the William and Flora Hewlett Foundation for the funding that has supported much of the teaching and learning reflected in this brief report. Thanks as well to our colleagues: Edward Cecil, Joseph Crocker, David Frossard, Kay Godel-Gengenbach, Joan Gosink, Jon Leydens, Ning Lu, Juan Lucena, Barbara Moskal, Junko Munakata-Marr, Jennifer Schneider, George (Jerry) Sherk, Marcelo Simões, Catherine Skokan, James Straker, Julie VanLaanen, and Sandra Woodson. Student readers who have made helpful comments on different versions include Josephine Cotton and Jill Savage.

Carl Mitcham and David Muñoz
June 2010

CHAPTER 1

Engineering

The first engineers in the United States, or at least the first to bear the title, were officers in the Revolutionary War; the first school of engineering here was a military academy, West Point. ... All our early engineering schools focused on mathematics, physics, chemistry, and drawing. ... There was little of the Latin, Greek, or Hebrew; classical literature; or rhetoric characteristic of the liberal arts college. ... [West Point and other] early engineering schools differed from the liberal arts colleges primarily in offering an education that was explicitly practical in a way that the college education of the day was not.

— Davis, M., 1998, *Thinking Like an Engineer*, pp. 18 and 20.

Before considering the possibility of humanitarian engineering, it will be useful to consider the character of engineering itself. The next chapter will address the issue of the meaning of humanitarianism. Only after reflecting on the meanings of engineering and of humanitarianism will Chapter 3 seek to imagine what we, along with others, call humanitarian engineering.

1.1 WHAT ENGINEERS DO

Engineers do engineering. Engineering, as commonly defined, is "the art of directing the great sources of power in nature for the use and the convenience of humans" (*McGraw-Hill Encyclopedia of Science and Technology*, 10th edition, 2007). Sometimes people debate whether this art (which entails creativity and innovation) includes science or the extent to which science is implicated in engineering. The encyclopedia entry from which the definition is quoted simply observes that engineering is differentiated from science because it directs "to useful and economical ends the natural phenomena which scientists discover and formulate into acceptable theories." But people can have different ideas about what counts as use, convenience, or useful and economic ends.

More directly than science, then, engineering is actively influenced by context. Unlike scientists, who claim to produce knowledge cut free from particular social settings, engineers aim to engage social settings by designing, constructing, and operating structures, machines, and diverse products, processes, and systems. Scientific knowledge is in principle context independent — although there are scholarly debates about whether this is really so or even possible (see, e.g., Biagioli, M., 1999). With engineering, however, there is less debate. Engineered products, processes, and systems are readily admitted to be context dependent. Engineers do not design and construct what they want but what others want — even if they cannot necessarily do this the way others sometimes want.

Engineers often have to educate clients about the limits of the physical universe as well as technical abilities. But in all these senses, engineering can only be engineering within some context.

Another way engineers often recognize their context dependency is to describe engineering as design under constraints. Such constraints may be physical, political, cultural, ethical, legal, environmental, or economic. But they are constraints presented to engineers, which then have to be incorporated into their designs. Instead, these are constraints with which engineers are presented, whereas they then have to incorporate into their designs.

On occasion engineers have claimed a history that goes back to the builders of medieval cathedrals, Roman aqueducts, or Egyptian pyramids. But these contexts — and constraints — were so different from those of the modern period that the only support for such a claim is to conceive engineering in quite abstract terms. Using a sufficiently abstract description, almost everything human beings do becomes a form of engineering. This is the argument, for instance, of the engineer philosopher Koen, B. (2003), when he identifies engineering with heuristic decision making. But insofar as engineering is a socially constructed profession with a contextualizing and constraining history continuous with the present, engineering arose in the late medieval or early modern period, initially in a military context.

1.2 FROM MILITARY TO CIVILIAN ENGINEERING

The term "engineer" is related to the Latin words *ingenium* (ingeniousness) and *ingeniatorum* (one who is ingenious). In Latin and other early modern language uses, the cognates have a double meaning of creative and devious. The mythical "genie in a bottle" has special powers to invent or make things, but it can also be a kind of demon.

The first persons explicitly denominated "engineers" were members of a military corps, those who designed and operated fortifications and various "engines of war" such as battering rams and catapults. The first institutions of engineering education were created by national governments and closely linked with the military, one early example being the Academy of Military Engineering established at Moscow in 1698 by Czar Peter the Great. What is often taken as the archetypal engineering school is the École Polytechnique, founded at Paris in 1794 to meet the needs of the Revolutionary Army of the Republic led by Napoleon Bonaparte. The Military Academy at West Point, founded in 1802, was the first institution of higher education to offer an engineering degree in the United States.

Engineers started to detach themselves from the military context during the Industrial Revolution in Great Britain. John Smeaton (1724-1792), a member of the Royal Society who began to use scientific methods to analyze construction projects, was the first to denominate himself a "civil engineer," which he did in 1768 on the title page of one of his reports. (The term was, however, already beginning to be used in the abstract form, "civil engineering," in other European languages.) It was Smeaton as well who established in 1771 the informal Society of Civil Engineers, which after his death came to be called the "Smeatonians." The Smeatonians influenced establishment in 1818 of the Institution of Civil Engineers (ICE), the first officially recognized professional engineering society. At the initial ICE meeting, one of the organizers described the engineer as "a mediator between the philosopher and the working mechanic and like an interpreter between two foreigners, [one who] must understand the language of both, [and hence possess] both practical and theoretical knowledge" (Institution of Civil Engineers, 1818).

Ten years later, when the ICE sought formal recognition by means of a royal charter, it asked one of its members, Thomas Tredgold (1788-1829), to draft a definition. His "Description of a Civil Engineer" begins:

Figure 1.1: Thomas Tredgold (1788–1829), a founding member of the Institution of Civil Engineers, London, who drafted the first clear definition of engineering.

> Civil Engineering is the art of directing the great Sources of Power in Nature for the use and convenience of man; being that practical application of the most important principles of natural Philosophy which has in a considerable degree realized the anticipations of [Francis] Bacon, and changed the aspect and state of affairs in the whole world. The most important object of Civil Engineering is to improve the means of production and of traffic in States, both for external and internal Trade. It is applied in the construction and management of Roads — Bridges — Raid Roads — Aqueducts — Canals — river navigation — Docks, and Storehouses for the convenience of internal intercourse and exchange; — and in the construction of Ports — Harbours — Moles breakwaters — and Light Houses, and in the navigation by artificial Power for the purposes of Commerce.

> Besides these great objects of individual & national interest it is applied to the protection of property where natural powers are sources of injury, as by embankments for the defence of tracts of country from the encroachment of the sea or the overflowing of Rivers; it also directs the means of applying Streams and rivers to use, either as powers to work machines, or as supplies for the use of Cities and Towns, or for irrigation; as well as the means of removing noxious accumulations, as by the drainage of Towns and Districts to prevent the formation of malaria, and secure the public health (Institution of Civil Engineers, Tredgold, T., 1828).

The initial sentence of this report was used to define civil engineering in the royal charter — and has ever since formed the core for definitions of engineering in general, as in that quoted from the *McGraw Hill Encyclopedia*.

Six features of the classic definition of engineering are worth highlighting. First, for Tredgold, civil engineering was simply all non-military engineering. In his description it clearly includes both what would now be called civil engineering (the designing of roads, dams, and related infrastructures), mechanical engineering (working with power and machines), and hydraulic engineering (irrigation and drainage). Thus, it is perfectly reasonable to apply his definition of civil engineering to engineering in general.

Second, engineering is an art. "Art," for Tredgold, means simply the skilled making of physical objects or artifacts. It is derived from the Latin *ars*, which translates the Greek *techne*. The original Greek does not imply fine art or aesthetics but indicates simply manual cunning or skill.

Third, what Tredgold termed "the great sources of power in nature" would today less rhetorically be called simply the resources of nature. And although Tredgold was thinking "great" in the sense of wind, water, fire (especially coal and steam), as opposed to animals and humans, the subsequent discoveries and harnessing of chemical, electrical, and nuclear energy would only enlarge his perspective, not alter it.

Fourth, as Tredgold's second sentence indicates, engineering makes use of "natural philosophy." Natural philosophy was an early term for what became modern natural science. A direct use of and reliance on science helps distinguish engineering from, for instance, craft making and architecture. The point may be emphasized by noting that Smeaton and many other engineers were active members of the Royal Society (founded 1660), the first professional society organized to promote Baconian science. But engineering uses science to make artifacts; it is an artful more than a cognitive use of science.

Fifth, unlike premodern science, the natural philosophy or science envisioned by Francis Bacon (1561-1626) aimed at the "conquest of nature" for the "relief of man's estate" (*The Advancement of Learning*, 1605). Another key feature of science in the distinctly modern sense is the use of mathematical modeling as developed by Galileo Galilei (1564-1642) and his successors as a means to this new end of conquest. Prior to the transformation of the sciences initiated by Bacon and others, the pursuit of knowledge was oriented much more toward a contemplative appreciation of the wonders of non-human nature and the moral disciplining of human nature. Bacon argued that

Figure 1.2: Francis Bacon (1561–1626), whose promotion of an active engagement with the world influenced the conception of engineering.

all moral questions had been answered by Christian revelation and that humans were thus now free, even obligated, to turn their attention to taking control of non-human nature and using it to practice the Christian virtue of charity by relieving suffering and glorifying God by exercising dominion. Nature was now to be seen as a source of power or, more generally, as a resource available for human exploitation. Tredgold explicitly allies engineering with such a Baconian perspective and program.

Finally, "use and convenience" is an almost technical term associated with the development of utilitarian philosophy in Great Britain during this same period. For instance, David Hume in his *Enquiry Concerning the Principles of Morals* (1751), observed that "a machine, a piece of furniture, a vestment, a house well contrived for use and conveniency" is for this very reason considered beautiful, and then continued to argue for use and convenience under the general heading of "utility" as the foundation of morals. For Tredgold and other members of the ICE (Institution of Civil Engineers, 2008), use and convenience are non-problematically assumed to be synonymous with meeting human needs through the advancement of industrial and commercial interests, a view quite common at the height of the Industrial Revolution — although one that, over the next century, became subject to social and philosophical criticism.

1.3 USE AND CONVENIENCE, EXTENDED AND CRITICIZED

How to operationalize "use and convenience" in the engineering profession has been the subject of engineering ethics. For roughly the first hundred years, until the early 1900s, the initial assumption was implicitly maintained within the profession. The engineering ability to re-design the world and

Figure 1.3: David Hume (1711–1776), whose empiricist and utilitarian notion of a "use and convenience" focused ethics was adopted by the professional engineering community.

its usefulness in increasing human productivity, in conjunction with industrial economic expansion, was always assumed, precisely because of such utility, to be good.

Additionally, when engineers began to formulate explicit codes of ethics they tended to emphasize collaboration with business and industrial interests. Primary examples were the codes of ethics of the American Institute of Electrical Engineers (AIEE, adopted 1912), of the American Society of Mechanical Engineers (ASME) and of the American Society of Civil Engineers (ASCE), both of which were adopted in 1914. Each of these three codes was less than a page in length and stressed that "the engineer should consider the protection of a client's or employer's interests his first professional obligation" (to quote the AIEE code) or required the engineer to act simply "as a faithful agent or trustee" (ASCE language).

> With regard to the ASCE, one scholar has noted how the first code of ethics adopted by the ASCE was intended to describe, rather than guide, the behavior of ASCE members. … Early codes of ethics were intended to document and publicize existing standards of behavior (largely for the benefit of potential employers), not to establish ideals toward which ASCE members might strive (Pfatteicher, S., 2003, p. 21).

As she further observes, this descriptive code also admonished members to be true to existing practice and "to be loyal to their clients, their fellow engineers, and their profession" (Pfatteicher, S., 2003, p. 29). Paradoxically, although one goal of this early code making was to enhance public recognition and a degree of autonomy, because of the pride of place given to business interests and company loyalty, the practical effect was to undermine independence as much as to promote it. In other words, professional engineering — insofar as it articulated loyalty as a primary value — tended to promote a kind of self-imposed tutelage to its most immediate employers.

The primary context within which engineering arose as a non-military profession in the early 1800s was the Industrial Revolution, especially in England. Engineering has continued ever since to be closely involved with nationalist technological and economic projects in an increasing number of countries. Many people — from engineers themselves to politicians — have argued that economic development if not civilization is synonymous with engineering achievement. Indeed, in the mid-20th century it became common for codes of ethics to present "public safety, health, and welfare" as primary ideals of engineering governing engineering practice. The current ICE code, for instance, lists this as the third of six rules of professional conduct (Institution of Civil Engineers, 2008). By the early 21st century this had become an almost universal commitment among professional engineering societies (see the model code of the World Federation of Engineering Organizations, sidebar). Within some constraining context, engineers were argued to be those professionals best able to help design and construct a world to realize public conceptions of safety, health, and welfare. (For one critical interpretation of this historical trajectory, see Mitcham, C., 1994.)

World Federation of Engineering Organizations (WFEO)

Model Code of Ethics: Practical Provision Ethics (as adopted 2001 Mitcham, C. (1994))

Professional engineers shall:

- hold paramount the safety, health and welfare of the public and the protection of both the natural and the built environment in accordance with the Principles of Sustainable Development;promote health and safety within the workplace;
- offer services, advise on or undertake engineering assignments only in areas of their competence and practice in a careful and diligent manner;
- act as faithful agents of their clients or employers, maintain confidentially and disclose conflicts of interest;
- keep themselves informed in order to maintain their competence, strive to advance the body of knowledge within which they practice and provide opportunities for the professional development of their subordinates and fellow practitioners;
- conduct themselves with fairness, and good faith towards clients, colleagues and others, give credit where it is due and accept, as well as give, honest and fair professional criticism;
- be aware of and ensure that clients and employers are made aware of societal and environmental consequences of actions or projects and endeavor to interpret engineering issues to the public in an objective and truthful manner;
- present clearly to employers and clients the possible consequences of overruling or disregarding of engineering decisions or judgment;

- report to their association and/or appropriate agencies any illegal or unethical engineering decisions or practices of engineers or others.

This was the vision of U.S. President Harry S. Truman, for instance, in the famous "point four" of his inaugural address on January 20, 1949. After committing the United States to, first, supporting the United Nations; second, economic recovery; and third, countering (Communist) aggression; Truman announced that

Figure 1.4: Harry S. Truman (1884–1972), at his 1949 inauguration, when he committed the United States to a foreign policy that promotes development.

> Fourth, we must embark on a bold new program for making the benefits of our scientific advances and industrial progress available for the improvement and growth of underdeveloped areas …
>
> The United States is pre-eminent among nations in the development of industrial and scientific techniques. The material resources which we can afford to use for the assistance of other peoples are limited. But our imponderable resources in technical knowledge are constantly growing and are inexhaustible.
>
> I believe that we should make available to peace-loving peoples the benefits of our store of technical knowledge in order to help them realize their aspirations for a better life …

> Our aim should be to help the free peoples of the world, through their own efforts, to produce more food, more clothing, more materials for housing, and more mechanical power to lighten their burdens.

Although the word "engineering" does not appear in Truman's speech, it clearly constituted a significant part of what he had in mind. Instead of a simple economic context, Truman was creating for engineering a new context, one having to do with a particular national view of international relations. Indeed, well before Truman gave the notion public expression, on the basis in part of just such ideas, colonialist powers had been exporting their engineering prowess to countries in Latin America, Africa, and Asia. Then, as part of their various rebellions against the imperialist control by countries in Europe, peoples in the colonies themselves had been attempting to jump start engineering education in the contexts of their own development. Truman's point four commitment was subsequently rationalized in political adviser Walt W. Rostow's *The Stages of Economic Growth: A Non-communist Manifesto* (Rostow, W., 1960). Again, although he did not explicitly reference the term, engineering plays a key role in Rostow's conceptual sequence from traditional societies and the preconditions for take-off to take-off, drive to maturity, and age of high mass consumption.

Yet late in the same century in which engineering emerged as a civilian profession, there also arose a number of movements critical of different aspects of nationalist industrialization, especially in its imperialist guises. The so-called "utopian socialism" of Robert Owen (1771-1858), the labor movement, and the revolutionary socialism of Karl Marx (1818-1883) are all cases in point. Humanitarianism is another instance; and as a criticism of some of the implications of nationalism and imperialism, it indirectly invites engineers to self-examination — and to consider the possibility of contexts alternative to those in which their professional practices have commonly been pursued.

The focus in this examination, then, will be on humanitarianism as a socio-historical movement that presents an alternative context of particular relevance to engineering understood as the artful drawing on science to direct the resources of nature for the use and the convenience of humans. In what follows we seek to explore how the humanitarian context might give rise to alternative ways to understand use and convenience and thus new opportunities in the engineering profession at the levels of both practice and education.

CHAPTER 2

Humanitarianism

> Theory cannot equip the mind with formulas for solving problems nor can it mark the narrow path on which the sole solution is supposed to lie by planting a hedge of principles on either side. But it can give the mind insight into the great mass of phenomena and their relationships, then leave it free to rise into the higher realms of action.
>
> — Carl von Clausewitz, *On War*, Book VIII, Chapter 6.

Despite what some might expect, it is not out of place to begin an examination of humanitarianism with a quotation from one of the great war theorists of the modern period. Not only did modern warfare give special impetus to the humanitarian movement, but the point made by Clausewitz applies perhaps even more to humanitarian action than to warfare. It is not possible to provide formulas to solve all problems.

Multiple connections with engineering may also be quickly indicated. As noted in Chapter 1, engineering has roots in military history; engineering theory likewise seldom leads directly to univocal solutions to engineering problems. Additionally, one of the most consistent, critical accounts of the real world in which civil engineering attempts to realize its ideal of protecting public safety, health, and welfare in the broadest senses of these terms is found in the humanitarian movement, which emerged in parallel with the engineering profession itself. It was only shortly after the crystallization of engineering as a profession in the early 1800s that a challenge to some of its contextual assumptions emerged later in the same century with what has become known as humanitarianism. In order to introduce humanitarianism as a movement relevant to engineering we will begin with some conceptual distinctions and then review its historical development.

2.1 HUMANITARIANISM VERSUS HUMANISM AND HUMAN RIGHTS

Like engineering, humanitarianism is a complex phenomenon with its own variegated history. One must, for instance, distinguish philosophical from theological humanitarianism. Theologically humanitarianism designates the Christian doctrine that affirms the humanity and denies the divinity of Jesus Christ. This is, in fact, the original usage of the word and not the one with which we are concerned.

Philosophical humanitarianism itself can nevertheless take on religious associations. In the mid-1800s, for instance, partisans of what was called the "religion of humanity" were often termed "humanitarians." Humanitarianism in this sense was to profess, as with the German philosopher

Johann Gottfried Herder (1744-1803), a commitment to the advancement or perfection of the human race.

A less religiously inclined philosophical humanitarianism must be distinguished from simple humanism. Although the two are related, they are not the same. Humanism can involve one or more of three distinct claims. One claim is that there is something about being human which transcends all particular forms of humanity. This is illustrated by William Shakespeare in the *Merchant of Venice* (written in the late 1500s) when the Jewish banker, Shylock, responds to those who see him as less than human by saying, "Hath not a Jew eyes? Hath not a Jew hands, organs, dimensions, senses, affections, passions! ... If you prick us do we not bleed?" (act III, scene 1). This Renaissance humanism challenges any tendency to limit moral sympathy by means of religion, race, class, or other human differences, and is a view of human life that found special expression in art and literature.

A second, stronger claim is that human beings are of unique or special significance in the world, with some versions going so far as to conceive humans as the only beings of significance. This is the view espoused, for instance, by the existentialist philosopher Jean-Paul Sartre (1905-1980) when he declares, "There is no other universe than a human universe, the universe of human subjectivity" (*L'Existentialisme est un humanisme*, antipenultimate paragraph). Humanism in this second sense is a metaphysical view of the human being as superior in reality to all other possible beings.

Still a third humanism is that of secular or scientific humanism, which may well affirm the special reality of being human but also sees this special reality as part of nature. Scientific humanism is a naturalistic humanism. In the words that open "Humanism and Its Aspirations" (2003 and sometimes referred to as "Humanist Manifesto III"), "Humanism is a progressive philosophy of life that, without supernaturalism, affirms our ability and responsibility to lead ethical lives of personal fulfillment that aspire to the greater good of humanity."

Although not incompatible with humanism in the second and third senses, humanitarianism as an ethical and political orientation or practice need not involve either a metaphysical elevation of human beings or a scientific naturalism. Its minimal foundation is something closer to the first claim, together with criticism of narrow forms of human community, especially nationalism. Indeed, one succinct definition describes humanitarianism as active sympathy or compassion for all humans in need. Efforts to aid people in crisis situations, whether caused by other humans or natural, are thus commonly seen as exemplars of humanitarian action. Humanitarianism is sometimes seen as synonymous with humanitarian aid work or humanitarian relief.

A commitment to humanitarian action should also be distinguished from a commitment to human rights, although again the two are easily associated. Humanitarianism typically involves an effort to alleviate human suffering by responding to human needs, but not necessarily on the basis of respect for individual human rights. Although human rights may be important, active response to human need out of compassion for human suffering is seen as more important. In a like manner, efforts to promote and respect rights need not — even though they often do — entail the alleviation of suffering. Although it may seem like too fine a distinction, it is one thing to argue against slavery because of the suffering it causes slaves, another to reject slavery as incompatible with basic natural or human rights.

Figure 2.1: Albert Schweitzer (1875–1965), whose life has often been appealed to as exemplifying the highest ideals of Christian and even more than Christian humanitarianism.

2.2 HUMANITARIAN UNIVERSALISM

The roots of the humanitarian criticism of restricted forms of community and the promotion of equity or equality among humans are many. One root, for instance, is the cosmopolitianism of Greek and Roman philosophy. In contrast to the idea that our primary social bonds are to our fellow citizens in a city-state or *polis* such as Athens or Syracuse, some ancient philosophers argued that the whole *cosmos* (Greek for physical universe) constituted a kind of *polis*, making all human beings members of a single community. Diogenes of Sinope (c.400s BCE), when asked his citizenship, is reported to have answered, "I am a citizen of the world" (*kosmopolitês* in Greek).

Another root is Christian missionary theology which, from St. Paul to Albert Schweitzer (1875-1965), has argued a supernatural version of universalism; insofar as all human beings are created by and equal in the sight of God, they are members of a common community with obligations to care for one another. The German medical missionary, Schweitzer, is often taken as a model Christian humanitarian because of the way he sacrificed achievement and fame as a theologian and as a classical musician in order to spend much of his life running a clinic in Gabon, Africa.

Finally, other roots are to be found in the moral principles of Enlightenment philosophy in both the empiricist and rationalist traditions. With regard to empiricism, not only did the Scottish philosopher David Hume (1711-1776) argue the central importance of use and convenience, but he defended sympathy as the foundational moral sentiment; this sentiment, expressible as benevolence and concerned especially to secure such basic goods as food, shelter, and social relationships not just for ourselves but for all is what makes utility or use and convenience pleasing and thus structures human behavior. From the tradition of rationalism, the German philosopher Immanuel Kant (1724-1804) argued for recognition of a categorical obligation to treat all humans as ends in themselves; historically such an imperative came to serve as ideal and support for such political revolutionary movements as socialism. The effort to create a transnational community of scientists who share a

Figure 2.2: Immanuel Kant (1724–1804), whose argument for an unqualified obligation to treat all humans beings as ends in themselves, become one basis for promoting human rights and cosmopolitanism.

common pursuit of knowledge and the labor movement organization of workers who are equally victims of economic oppression have been further influences on the humanitarian movement.

None of these historical influences, however, adopted the term "humanitarianism." Indeed, in its initial secular uses in the early 1800s the term was largely derogatory, as denoting excess in the promotion of humane principles over more realistic or patriotic ones. People more concerned about the poor in a foreign country than the welfare of their own families were sometimes disparaged as "humanitarians." In the late 1800s, however, the term began to take on positive connotations, as when the American sociologist Lester F. Ward described humanitarianism as aiming "at the reorganization of society, so that all shall possess equal advantages for gaining a livelihood and contributing to the [common] welfare" (Ward, L., 1883, p. 450). Only after the fact were social movements grounded in active compassion directed toward meeting the basic needs of all persons irrespective of national or other distinctions — often with a special focus on health care, food, and shelter — interpreted as expressions of something called a humanitarian movement.

2.3 ANTICIPATIONS OF THE HUMANITARIAN MOVEMENT

The social reform movements of the 1700s and 1800s that can most easily be interpreted as anticipations of humanitarianism proper generally criticized restricted forms of human community such as racism, class distinctions, and discrimination against women, along with mistreatment of the mentally ill or handicapped and prisoners especially insofar as they limited access to basic needs. Eventually active compassion was extended even to animals; the humane treatment of animals requires that they too not be deprived of food or shelter and treated with some minimal level of benevolence.

Humanitarianism as an ethical vision was also closely associated with the creation of the social sciences. During the 19th century, modern science began to explore social phenomena, in part to deal with the conditions created by new human powers over the natural world. Initially in England, then in France, and later in the United States, centers of raw materials extraction and industrial production gave rise to a working class living in deplorable conditions. Industrial technologies created urban centers that needed better management for the benefit of the workers who lived in them — not because they were members of some political, religious, economic, or ethnic group, but simply as human beings, who could also be scientifically studied as such. Public health and public engineering were for the benefit of all, although the "all" was in the first instance understood within a national context.

Humanitarianism thus aimed to extend compassion beyond traditional family, class, or village limits, especially through the utilization of science broadly construed. Although this may appear to have been simply a secular version of Christian missionary work — especially since humanitarian organizations often attracted voluntary contributions from believers — the increasing number of middle-class persons involved in providing relief for the victims of warfare and the improvement of urban slums constituted a historically unique social movement.

The first major anticipation of that active compassion that would come to be called humanitarianism addressed the issue of slavery. Until the 1700s there was little intellectual or popular opposition to the theory and practice of human bondage. The Greek philosopher Aristotle had argued that some humans were "by nature" subject to other humans; and St. Paul in his "Letter to Philemon" simply accepts the social institution of slavery, counseling slaves to please their masters and masters to be kind to slaves.

It is sometimes thought that the abolitionist movement emerged solely from enlightenment philosophy and radical forms of Christianity such as Quakerism. Prior to the 18th century, however,

Christian theology did exercise some restricting influence. Charles V of Spain, for instance, as a result of interventions by the missionary Dominican monk Bartolomé de las Casas, in 1542 freed all native American slaves and made them citizens of the Spanish Empire (although the importation of African slaves continued). (For an introduction to the complex history of slavery, see, e.g., Peabody, S., 2004.)

Two hundred years later the movement to abolish the enslavement of persons of African descent began in England and rather quickly achieved significant political results. The politician William Wilberforce (1759-1833), after an evangelical experience, in 1787 brought the abolitionist campaign to Parliament. A decade later, in 1794 the new republic established by the French Revolution abolished slavery in France and its colonies. Then twenty years after he began, in 1807, Wilberforce succeeded in securing a majority vote to outlaw the slave trade throughout the British Empire; in 1833 the British Parliament went further and abolished slavery itself. Within another three decades the American Civil War (1861-1865) had abolished African slavery in the United States, and by the end of the century the same had occurred in Brazil and all other countries of the western hemisphere. The struggle for racial equality has been a key anticipation of humanitarianism universalism.

A second anticipation of humanitarianism is associated with the promotion of child welfare and labor protection legislation, and involved as well a democratic extension of the voting franchise that in effect challenged class privilege and economic discriminations. Until the mid-1800s all European and American democracies had property qualifications for voting rights. The gradual relaxation and eventual elimination of these class qualifications went hand in hand with efforts to meliorate the condition of workers — especially women and children — during the Industrial Revolution (c.1750-1850). The utilitarian philosophy of Jeremy Bentham (1748-1832) and the socialist activism of Robert Owen (1771-1858) were both founded on a heightened sympathy for the oppression of wage laborers. According to one historian, extension of the voting franchise, the emancipation of women, and labor laws were all first and foremost manifestations of "the rise if a new spirit of humanitarianism" in the 1800s. The "humane spirit that sees clearly enough and feels keenly enough the wrongs of the lowly and disinherited to make strenuous efforts to redress those wrongs" is as distinguishing a feature of the late 19th and early 20th centuries as "the progress of invention or the spread of elementary education" (Smith, P., 1957, vol. 2, p. 588).

The emancipation of women and woman's suffrage may be identified as a third distinctive manifestation of proto-humanitarianism. Although a few countries allowed women to vote earlier, and sometimes on the basis of a humanitarian-like appeal, it was not until the early decades of the 1900s that woman's suffrage was widely accepted — more often on the basis of human rights arguments than humanitarian ones.

The difference between active compassion humanitarianism and human rights is further illustrated by efforts to reform the treatment of criminals, the physically handicapped, and the mentally ill. Italian philosopher Cesare Beccaria (1738-1794) argued forcefully against torture and invented the science of penology on the basis of sympathy for the excessive sufferings of those in prisons.

Transformations also began to take place in the treatment and care of the handicapped and mentally ill — almost always on the basis of sympathy and seldom because of a concern for rights. When concern also began to be expressed for the humane treatment of animals, it was again often initially justified on the basis of a desire to reduce suffering rather than respect rights.

For present purposes, however, humanitarianism will be restricted to a particular movement that has used this term to define itself. The history of self-identified humanitarian movement can be outlined in five overlapping phases in the development of an active compassion for the basic needs of all persons irrespective of national or other distinctions. Initially there was a special focus on immediate medical care for those suffering from warfare, but this was subsequently extended into concerns for health care more generally as well as needs for water, food, and shelter of anyone suffering from disaster, whether human-caused or natural. Given the fact that those in need will for all practical purposes be those lacking status or power, humanitarianism may conveniently be summarized as action compassion directed toward meeting the basic needs of all — especially the powerless, poor, or otherwise marginalized.

2.4 PHASE ONE (1800S): RISE OF THE HUMANITARIAN MOVEMENT PROPER

With regard to the first phase, the humanitarian movement is generally understood to have originated in the mid- to late 1800s. This origination is associated with the rise of the profession of nursing, as promoted in the work of Mary Seacole (1805-1881) and Florence Nightingale (1820-1910) in the Crimean War (1854-1856) and Clara Barton (1821-1912) in the U.S. Civil War (1861-1865). But the key event was the reaction of Swiss businessman Henri Dunant (1828-1910) to the Battle of Solferino (1859), which ended the Second Italian War of Independence.

In nine hours of fighting, the Battle of Solferino resulted in approximately 30,000 Austrian, Italian, and allied French casualties. When Dunant, on a business trip, accidentally witnessed the industrial carnage of the Solferino battlefield and the tendency of medical personnel from each army to restrict attention to their own injured, he was stimulated to imagine a new kind of medical care that would address need irrespective of national identity. This vision led to the 1863 creation of the International Committee of the Red Cross/Red Crescent (ICRC), which currently defines itself as "an impartial, neutral and independent organization whose exclusively humanitarian mission is to protect the lives and dignity of victims of armed conflict and other situations of violence and to provide them with assistance." Additionally, as it states on its web site (`icrc.org`) the ICRC "endeavours to prevent suffering by promoting and strengthening humanitarian law and universal humanitarian principles."

Major tests for the new organization and its commitment to ameliorating the conditions of wounded and displaced peoples were presented by the Franco-Prussian War (1870-1971), the Spanish-American War (1898), and a number of natural disasters, with its successes leading to Dunant receiving the first Nobel Peace Prize (1901). The ICRC itself subsequently received its own

Figure 2.3: Mary Seacole (1805–1881), Florence Nightingale (1820–1910), and Clara Barton (1821–1912), three founders of the profession of nursing as one of the original expressions of humanitarian movement.

Nobel Peace prizes in 1917 (for work in World War I), 1944 (for work in World War II), and 1963 (on the occasion of its 100th anniversary).

The progressive institutionalization of the ICRC involved creation of the Geneva Accords for the conduct of hostilities and granted battlefield protection to all medical personnel. In the course of the formulation and promotion of such accords, the ICRC has made claim to being the custodian of international humanitarian law, the development of which has become a major feature of international relations or relations among states since the early 20th century. In conjunction with such developments, the ICRC also presented itself as a neutral institution attending to the needs of prisoners of war, but on condition that it respected national sovereignty and only reported any observed mistreatment of prisoners to the offending state, not to other states or the international community.

2.5 PHASE TWO (EARLY 1900S): HUMANITARIANISM BEYOND THE BATTLEFIELD

During a second phase, the first half of the 20th century saw the development of new forms of humanitarianism that expanded the movement beyond the limits of medical care directed toward military personnel. The ICRC became concerned with the plight of civilian non-combatants and for persons caught in natural disasters. New models of humanitarianism can be found in the work of Norwegian scientist and explorer Fridtjof Nansen (1861-1930) and of U.S. mining and civil engineer Herbert Hoover (1874-1962): Nansen in post-World War I work resettling refugees under

Figure 2.4: Henri Dunant (1828–1910), founder of the International Committee of the Red Cross/Red Crescent (IRCRC), which is generally recognized as the senior institutionalization of the humanitarian movement.

the auspices of the League of Nations, and Hoover in relief work during and after the war as well as in response to the Great Mississippi Flood of 1927.

This period also witnessed the emergence of humanitarian NGOs other than the ICRC: e.g., Baptist World Aid (1905), American Friends Service Committee (1917), Catholic Medical Mission Board (1928), Save the Children (1932), OXFAM (1942), and CARE (Cooperative Action for American Relief Everywhere, 1945). With regard to the ICRC, its knowledge of the Holocaust, war crimes, and crimes against humanity during World War II — and its principled refusal to reveal these to the world because of its respect for national sovereignty — raised fundamental questions about some its operating assumptions. Creation of the United Nations (1945) and international adoption of the Universal Declaration of Human Rights (1948) provided a further basis for questioning the primacy of national sovereignty.

2.6 PHASE THREE (1950S-1960S): HUMANITARIANISM AS FREE WORLD IDEOLOGY

In a third phase, however, something like humanitarian development became a kind of free-world ideological alternative to Communism. This was the explicit proposal of Truman and was embodied as well in the European Recovery Program or Marshall Plan (1947-1951). The creation of international agencies such as the United Nations Educational, Scientific and Cultural Organization (UNESCO, 1945), the UN International Children's Emergency Fund (UNICEF, 1946), the UN High Commission for Refugees (UNHCR, 1950), the Organization for Economic Cooperation and Development (OECD, 1961), the U.S. Peace Corps (1961), the World Food Programme (WPG,

1963) — together with a series of UN peacekeeping actions (India-Pakistan, 1949; Suez, 1956; Congo, 1960; et al.) — combined to give humanitarianism the character of a anti-communist program. (As an aside, the ICRC, UNHCR, UNICEF, and WFP are sometimes thought of as the "big four" humanitarian relief agencies.)

Insofar as it grew out of post-World War II relief and recovery efforts, this third phase in the historical development of humanitarian thinking also highlighted efforts that go beyond some immediate response to a crisis. Simple crisis intervention humanitarianism, it was increasingly recognized, needs to be complemented with crisis recovery humanitarianism.

During this period as well the ICRC initiated an act of self-reflection that led to the formulation of seven key principles (see sidebar). As one historian has summarized the situation,

> For a long time the ICRC acted to help individuals in conflicts but without any official principles, much less doctrine or general policies. The emphasis was on pragmatic moral accomplishments — even if there was some attention to legal rules and precedent (Forsythe, D., 2005, p. 161).

The seven key principles are humanity, impartiality, neutrality, independence, unity, universalism, and volunteerism. But as the same historian notes, not all these principles are equal. "Impartiality, neutrality, and independence are essential means to the central goal of ICRC humanitarian protection" (Forsythe, D., 2005, p. 161). The remaining three principles simply define operational relations between the ICRC and various national organizations, and the last — volunteerism — is highly qualified by the need for considerable full-time staff.

2.7 PHASE FOUR (1970S-1990S): ALTERNATIVE HUMANITARIANISMS

Beginning in the late 1960s, however, and indicative of a fourth phase, humanitarianism began to separate itself from its previous close association with anti-communism. One key event was the Nigerian Civil War in the break-away province of Biafra (1969), which became as well the first televised international humanitarian crisis. The experience within the disaster relief community was one of gut wrenching paradox: providing relief that only enabled killing to continue and become more murderous.

International Committee of the Red Cross and Red Crescent

The Fundamental Principles

Extract from XXVIth International Conference of the Red Cross and Red Crescent

Humanity

The International Red Cross and Red Crescent Movement, born of a desire to bring assistance without discrimination to the wounded on the battlefield, endeavours, in its international and national capacity, to prevent and alleviate human suffering wherever it may be found. Its purpose is to protect life and health and to ensure respect for the human being. It promotes mutual understanding, friendship, cooperation and lasting peace among all peoples.

Impartiality

It makes no discrimination as to nationality, race, religious beliefs, class or political opinions. It endeavours to relieve the suffering of individuals, being guided solely by their needs, and to give priority to the most urgent cases of distress.

Neutrality

In order to continue to enjoy the confidence of all, the Movement may not take sides in hostilities or engage at any time in controversies of a political, racial, religious or ideological nature.

Independence

The Movement is independent. The National Societies, while auxiliaries in the humanitarian services of their Governments and subject to the laws of their respective countries, must always maintain their autonomy so that they may be able at all times to act in accordance with the principles of the Movement.

Voluntary Service

It is a voluntary relief movement not prompted in any manner by desire for gain.

Unity

There can be only one Red Cross or Red Crescent Society in any one country. It must be open to all. It must carry on its humanitarian work throughout its territory.

Universality

The International Red Cross and Red Crescent Movement, in which all societies have equal status and share equal responsibilities and duties in helping each other, is worldwide.

Under such conditions, humanitarian aid workers began to challenge even more strongly than had been done after World War II the principle of respect for national sovereignty. Aid workers began to want to openly criticize governments on both sides of the civil war and governments outside the conflict supporting one side or the other. The resulting crisis of conscience in the humanitarian community catalyzed the founding, by the French physician Bernard Kouchner, of *Médecins sans Frontieres* (MSF or Doctors without Borders) in 1971. MSF, which has become the largest non-governmental relief agency in the world, grew out of dissatisfaction with the inability of the Red Cross/Crescent to react independently of national government controls, and its tendency to remain within safe boundaries; it refused to be limited by state sovereignty. The idealistic physicians of MSF pioneered new ways to bring medical care to people in crisis and to speak out against human rights abuses by state and non-state actors alike. Over the course of what came to be known as the "decade of the refugee" (1975-1985), as people fled state-initiated disasters in a series of countries from Indochina to Africa and Afghanistan, respect for the sovereignty of states that were in fact killing their peoples became increasingly hard to defend, and MSF has responded to needs resulting from earthquakes, hurricanes, war, and famine in Central America, Africa, Russia, the Balkans, and the Middle East (Tanguy, 1999).

2.8 PHASE FIVE (2000S-PRESENT): HUMANITARIANISM GLOBALIZED AND QUESTIONED

Finally, in the context of the end of the Cold War (early 1990s) there arose two quite different trajectories in humanitarianism. The first, more widely adopted trajectory, has been what may be called the globalization of humanitarianism. This trajectory is best represented by the "United Nations Millennium Declaration" (2000), in which the member states recognized, "in addition to separate responsibilities to [their] individual societies, ... a collective responsibility to uphold the principles of human dignity" and a duty "to all the world's people, especially the most vulnerable" (Section I, paragraph 2). In addition, the national signatories affirmed a belief that

> the central challenge we face today is to ensure that globalization becomes a positive force for all the world's people. ... [O]nly through broad and sustained efforts to create a shared future, based upon our common humanity in all its diversity, can globalization be made fully inclusive and equitable. These efforts must include policies and measures, at the global level, which correspond to the needs of developing countries and economies in transition and are formulated and implemented with their effective participation (Section I, paragraph 5).

The "Millennium Declaration" was extended into the Millennium Project, commissioned by UN Secretary-General Kofi Anan in 2002 to develop a concrete action plan to eradicate the most extreme poverty by 2015. In this project humanitarian action came to focus not so much on crisis relief or even recovery as on crisis prevention humanitarianism. (See Millennium Development Goals sidebar.) Stress nevertheless remained on working to meet basic human needs for food and shelter, with a special emphasis being placed on water, from the providing of potable water to the treatment of waste water. The millennium project aim has been further argued by economist Jeffrey Sachs, head of an advisory group for the Millennium Project, in his book *The End of Poverty* (2005).

Millennium Development Goals

The eight Millennium Development Goals (MDGs) constitute an effort to operationalize the United Nations Millennium Declaration (September 2000). The MDGs (adopted in 2001) are

1. Eradicate extreme poverty and hunger
2. Achieve universal primary education
3. Promote gender equality and empower women
4. Reduce child mortality
5. Improve maternal health
6. Combat HIV/AIDS, malaria, and other diseases
7. Ensure environmental sustainability
8. Develop a global partnership for development

A second, more skeptical trajectory of thought is associated with the persistence and even intensification of various strident forms of nationalism and the rise of non-state actor terrorism. From this perspective, the crisis of humanitarianism has become only more acute. Indeed, worry that ethnic-based nationalisms could lead to indefinite warfare was part of a European resistance to military action led by the United States in the Balkans. Additionally, the "humanitarian war" against Yugoslavia during the Kosovo campaign (1999) called into question the whole meaning of humanitarianism, as has the so-called "war on terror" that became a central feature of international relations post-2001.

Two perspicacious articulations of the early 21st century crisis in humanitarianism can be found in observations by David Rieff and David Kennedy, two persons who have been deeply involved with the humanitarian movement. As Rieff observes,

> [H]umanitarianism is an impossible enterprise. Here is a saving idea that, in the end, cannot save but can only alleviate. ... For there are, as Sadako Orgata, the former head of UNHCR, put it, "no humanitarian solutions to humanitarian problems." More than that, the pressures on humanitarian workers ... have become all but intolerable (Rieff, D., 2002, p. 86).

Or, in the words of Kennedy,

> We promise more than can be delivered — and come to believe our own promises. We enchant our tools. ... At worst, ... our own work [contributes] to the very problems we hoped to solve. Humanitarianism tempts us to hubris, to an idolatry about our intentions

and routines, to the conviction that we know more than we do about what justice can be (Kennedy, D., 2004, p. xviii).

At the same time, in spite of these well recognized difficulties and failings, it is important to acknowledge that there have been many successes. As other observers have argued, humanitarian action has made important contributions to the lives of many of the powerless and poor (see, e.g., DiPrizio, R., 2002; Minear, L., 2002; Terry, F., 2002, and Architecture for Humanity, 2006). It is important not to let the perfect — including the absence of a perfect theory — be the enemy of the good. To recall again the counsel of Clausewitz, "Theory cannot equip the mind with formulas for solving problems nor can it mark the narrow path on which the sole solution is supposed to lie [although] it can give the mind insight into the great mass of phenomena and their relationships."

2.9 THE HUMANITARIAN CHARTER

In order to provide the humanitarian mind insight and orientation amid the great mass of phenomena and relationships with which it must deal — especially in a situation of globalized questioning — in 1997 a group of humanitarian NGOs and the Red Cross and Red Crescent movement undertook to draft a Humanitarian Charter and to identify Minimum Standards for disaster assistance. The group and its effort became known as "The Sphere Project." A "trial edition" of the what is usually referred to as the Sphere Handbook, officially titled *Humanitarian Charter and Minimum Standards in Disaster Response*, was published in 1998, with the "first final edition" appearing in 2000 followed by a second edition in 2004.

The Handbook is a volume of more than 300 pages that includes a short Charter, statement of common standards, and chapter elaborations of minimum standards, key indicators, and guidance notes in four areas:

- Water supply, sanitation and hygiene promotion;
- Food security, nutrition and food aid;
- Shelter, settlement and non-food items; and
- Health services.

This extensive and detailed material is followed with a clearly articulated "Code of Conduct" for ICRC and other NGOs (see sidebar). The Handbook as a whole "is designed for use in disaster response, and may also be useful in disaster preparedness and humanitarian advocacy" (Sphere Project, 2004, p. 6).

The Code of Conduct

(adapted from the Sphere Project, 2004)

Principles of Conduct for The International Red Cross and Red Crescent Movement and NGOs in Disaster Response Programs

1. The humanitarian imperative comes first.
2. Aid is given regardless of the race, creed or nationality of the recipients and without adverse distinction of any kind. Aid priorities are calculated on the basis of need alone.
3. Aid will not be used to further a particular political or religious standpoint.
4. We shall endeavor not to act as instruments of government foreign policy.
5. We shall respect culture and customs.
6. We shall attempt to build disaster response on local capacities.
7. Ways shall be found to involve program beneficiaries in the management of relief aid.
8. Relief aid must strive to reduce future vulnerabilities to disaster as well as meeting basic needs.
9. We hold ourselves accountable to both those we seek to assist and those from whom we accept resources.
10. In our information, publicity and advertizing activities, we shall recognize disaster victims as dignified humans, not hopeless objects.

According to the Handbook, Sphere is grounded in two fundamental beliefs: "first, that all possible steps should be taken to alleviate human suffering arising out of calamity and conflict, and second, that those affected by disaster have a right to life with dignity and therefore a right to assistance" (Sphere Project, 2004, p. 5). Thus, does the humanitarian movement itself understand the meaning of action compassion for the basic needs of all — especially the powerless, poor, or otherwise marginalized. The basic needs of suffering and dispossessed must not just be met; they must be met with a respect for their dignity and human rights.

The four-page "Humanitarian Charter" outlines three principles: (1) the right to life with dignity, (2) the distinction between combatants and non-combatants, and (3) *non-refoulement* or the principle that "no refugee shall be send (back) to a country in which his or her life or freedom would be threatened." Although the Charter admits that principle two has become increasingly problematic in practice, it nevertheless argues for struggling to maintain it whenever possible.

The three principles are supplemented with five paragraphs on roles and responsibilities that further recognize the complexities of contemporary humanitarian relief work. Although the state has the primary responsibility to protect and assist all its citizens, "those with primary responsibility are not always able or willing to perform this role themselves," which is what occasions humanitarian intervention. Furthermore, often when warring parties fail to respect humanitarian interventions, such interventions can actually enhance the vulnerabilities of civilians. Humanitarian workers are "committed to minimizing any such adverse effects." Additionally, humanitarians are committed to "the protection and assistance mandates of the [ICRC] and of the United Nations High Commissioner for Refugees under international law."

Finally, the Sphere Handbook admits that the Charter and associated standards "will not solve all of the problems of humanitarian response." But in words that echo Clausewitz's observation with regard to theory, the Handbook suggests that it can serve as "a tool for humanitarian agencies to enhance the effectiveness and quality of their assistance, and thus to make a significant difference to the lives of people affected by disaster" (Sphere Project, 2004, p. 14). The Charter constitutes a good faith effort to maintain the humanitarian ideal in the face of a manifold of complexities.

CHAPTER 3

Humanitarian Engineering

Technology influences disaster aid in two ways: First, it enables agencies to devise quick, although not necessarily appropriate, solutions to needs. Second, it influences the way in which needs are perceived and thus indirectly shapes many approaches.

— Cuny, F., 1983, *Disasters and Development*, pp. 138–139.

It is against the dual background of engineering as context dependent (Chapter 1) and the new context of humanitarianism (explored in Chapter 2) that what is called "humanitarian engineering" has emerged. In general terms, engineering is the *artful drawing on science to direct the resources of nature for the use and the convenience of humans*. Humanitarianism has been generalized as an *active compassion directed toward meeting the basic needs of all — especially the powerless, poor, or otherwise marginalized*. Humanitarian engineering may thus be described as the *artful drawing on science to direct the resources of nature with active compassion to meet the basic needs of all — especially the powerless, poor, or otherwise marginalized*

The concept of humanitarian engineering has been independently developed in other contexts as well. For instance, a graduate student in mechanical engineering at Queen's University, Canada, did an extended analysis of humanitarian engineering in the engineering curriculum using the definition of humanitarian engineering as "the application of engineering skills specifically for meeting the basic needs of all people, while at the same time promoting human (societal and cultural) development" (VanderSteen, J., 2008, p. 8). The close similarities to our proposed working definition should be obvious.

To some degree humanitarian engineering is related to what Mitcham, C. (2003) has termed "idealistic activism" among scientists and engineers, as exemplified by organizations such as International Pugwash (founded 1957) and the Union of Concerned Scientists (founded 1969). Among a diverse collection of related organizations seeking to build bridges between humanitarianism and scientific technology are the Responsible Care initiative of the American Chemistry Council and the International Network of Engineers and Scientists for Global Responsibility (INES). Responsible Care, founded in 1988, is a voluntary program to improve environmental health and safety in the chemical and related industries, especially in developing countries. INES, founded at a 1991 international congress in Berlin, is an association of more than 90 organizations in 50 countries promoting the involvement of technical professionals in humanitarian and peace development activities. But arguably the most significant and influential figure on the emergence of humanitarian engineering as such was the civil engineer Frederick (Fred) Cuny (1944-1995). It is thus appropriate to begin a discussion of humanitarian engineering with the Fred Cuny story.

Figure 3.1: Fred Cuny (1944-1995), whose humanitarian engineering work has become an inspiration to many aspiring humanitarian engineers.
(Source: `www.world.std.com/~jlr/doom/cuny.htm` Credit: Judy DeHass).

3.1 THE FRED CUNY STORY

Versions of the Cuny story are available in at least three formats. A Public Broadcasting System *Frontline* program, *The Lost American* (1997), told his story in an investigative television report. Michael Pritchard's "Professional Responsibility: Focusing on the Exemplary" (1998) is an academic argument inspired in part by Cuny. Scott Anderson's *The Man Who Tried to Save the World* (Anderson, S., 1999) is a book-length biography. What follows draws on these and other sources, but simplifies.

Fred Cuny was born in New Haven, Connecticut, but in 1952 his family moved to Texas where, after graduating from high school, he studied civil engineering at Texas A&M University. As a student, he was fascinated with flying airplanes and wanted for a time to become a pilot in the U.S. Marines. He also took some courses dealing with urban planning and development and became increasingly involved with the problems of Mexican migrant workers in south Texas. After doing a small amount of engineering for the new Dallas-Ft. Worth international airport (begun in

1966), Cuny went in search of a more fulfilling form of engineering work. He wound up flying relief supplies for CARE into the civil war between Nigeria and the break away region of Biafra (1967-1970). Afterward he returned to Dallas to found the Intertect Relief and Reconstruction Corporation and became involved in a series of disaster relief operations.

Two early involvements were with the natural disasters caused by earthquakes in Nicaragua in 1971 and in Guatemala in 1976. His experience there led to formulation of what became known as the "Cuny approach" to disasters, using them as catalysts to improve people's lives instead of simply working to return to the status quo. As Cuny observed in *Disasters and Development* (Cuny, F., 1983), an influential analysis sponsored by Oxfam America and summarizing some of the lessons learned from these experiences, in the past there had been a tendency for disaster relief and development workers to go their separate ways. Disasters were seen as one thing, development another. But because developing countries were more likely to experience disasters, and the experience could undermine momentum toward development, Cuny argued for new forms of collaboration. Disasters, Cuny thought, should be seen as opportunities to promote development.

> Generally, speaking, simply helping victims until they can get going [that is, disaster relief work] has little overall impact on reducing recovery time and, depending on how the aid is provided, may even prolong it. ... Helping people *to* recover [a form of development work], ... can demonstrably reduce recovery time (Cuny, F., 1983, p. 202).

As he summarized this insight, "disasters can be a primary cause of underdevelopment, as well as intertwined with a country's progress toward development" (Cuny, F., 1983, p. 206).

Another of Cuny's key beliefs was that there was a distinct place for engineering in humanitarian disaster relief work. As he pointed out in his analysis of the Guatemalan case,

> none of the [disaster relief] agencies had contact with any of the major earthquake engineering institutions, nor (with only one or two exceptions) were the institutions consulted during the course of the reconstruction efforts. The earthquake engineering institutions, for their part, completely ignored the program implementors. Several international earthquake engineering organizations sent survey teams to Guatemala to study the collapse of buildings, bridges, and other structures, but made virtually no contact with those organizations involved in reconstruction programs (Cuny, F., 1983, pp. 134–135).

From Cuny's perspective, even in disaster relief more than physicians and nurses were needed. Likewise in the work of development, there exists a need for more than agricultural specialists and agronomists. In both disaster and development work, engineering knowledge and skills have been under-utilized. Indeed, it was precisely through the use of his own skills in engineering analysis and design that Cuny came to think about disaster and development in new ways and to reach conclusions somewhat at odds with existing assumptions.

One of these conclusions concerned the importance of not waiting for disasters to happen, but in planning for them. Planning and the development of strategies to deal with possible disasters such as earthquakes and hurricanes is not as glamorous or adrenalin producing as jumping into a disaster. In Cuny's words, "Typical preparedness activities include predetermination of effective strategies and appropriate modes of involvement, development of tools needed by the emergency staff, development of plans for the actual response, and training for crisis operations." This kind of work is "normally seen as an activity of the planning and engineering disciplines," not just of medical or military personnel (Cuny, F., 1983, p. 220). Planning requires analysis and office work, if you will. It can also save lives more effectively than simply waiting for disaster to strike. Earthquake or other natural disaster preparedness means that when such events occur there will be less disruption in peoples' lives and recovery will be much quicker. In a subsequent analysis of famines, Cuny and Hill (1999) likewise argued for the priority of counter-famine planning and preemptive strategies over conventional famine relief.

In some ways Cuny picked up and extended aspects of the alternative technology movement from the 1960s and ideas such as those associated with E.F. Schumacher's *Small Is Beautiful* (Schumacher, E., 1973). Although Cuny promoted the importance of engineering and technology, for instance, he also recognized their limits.

> In the future, not only will the scope of the disaster problem increase, but so will the diversity of the challenges. As both our technological capabilities and our understanding of the social and economic aspects expand, we will be called upon to participate in a wider range of activities than ever before. The technological aspects will bring, perhaps, the most visible changes. ... [But as] disaster technologies and capabilities are improved, [volunteer organizations] will be faced with a parallel challenge — how to keep technology at an appropriate level (Cuny, F., 1983, pp. 258–259).

Following his work in Central America, Cuny became involved in disaster and development work during the Sudan-Ethiopia famine (1985) and with the Kurds in Iraq (1991); during the Somalia relief operation (1992); and to repair the water system during the siege of Sarajevo (1993-1994). His Sarejevo work lead to a "Talk of the Town" report and interview in *The New Yorker* (November 22, 1993). As an anonymous author wrote, "Over the years, Cuny and his associates have honed an approach that honors solid technical competence over vaporous good intentions." But why should "doctors and medicines routinely get flown in [to disaster areas], rather than engineers and piping"? The biggest problems in Sarejevo, according to Cuny's engineering analysis,

> were water and heat and, to a lesser degree, food, and, specifically, the way people were having to leave the relative security of their homes to forage and line up for such necessities for hours at a time, thereby subjecting themselves to the relentless bombardments and sniper fire. The key, therefore, lay in finding, as quickly and efficiently as possible, some way of getting such things to people in their homes, so that they wouldn't have to always be going out for them.

This Cuny did by engineering a new water system and providing the tools and instructions so that people could safely tap into a gas pipeline that was available. Then to help out with the monotony of United Nations food supplies, he distributed seeds for potted gardens on apartment balconies.

In 1995 Cuny was awarded a MacArthur Foundation Fellowship in a program designed to recognize "hard-working experts who often push the boundaries of their fields in ways that others will follow." But at the time of the award Cuny was on a mission to broker a peace agreement in Chechnya. He was assassinated there before the award notification was able to reach him.

3.2 OTHER PRECURSORS AND INFLUENCES

Cuny's life and work may have exercised a particularly strong influence on the humanitarian engineering ideal — his story is especially useful in helping students imagine new possibilities for engineering careers — but there were other precursors and influences as well. Two engineers who became U.S. Presidents, the engineer co-founder of the United States Peace Corps, and the NGO Doctors without Borders are all cases in point.

Herbert Hoover (1874-1964), who earned a degree in mining engineering from the first class of Stanford University, became the 31st U.S. President (1929-1933) primarily on the basis of his success as the organizer of European relief work during and immediately after World War I (1914-1918) and then again as the first Secretary of Commerce (under Presidents Warren Harding and Calvin Coolidge) especially during the Great Mississippi Flood of 1927. Although disaster relief was not part of the remit of the Secretary of Commerce, the governors of affected states requested that Hoover lead an effort that became a model for mobilizing and coordinating volunteers, local authorities, and national agencies. With a special grant from the Rockefeller Foundation, Hoover also set up health units that helped address problems of malaria, pellagra, and typhoid fever throughout the region. From Hoover's own perspective his success was grounded in his training as an engineer and his commitment to the promotion of efficiency through the use of analytic experts to identify problems and then propose solutions.

Jimmy Carter (born 1924), the 39th President (1977-1981), graduated from the United States Naval Academy at Annapolis with a BS in engineering. After his term as president he became a well known humanitarian working, for instance, with Habit for Humanity and home construction that could be presumed to draw on his engineering skills. Carter was awarded the Nobel Peace Prize (2002) "for his decades of untiring effort to find peaceful solutions to international conflicts, to advance democracy and human rights, and to promote economic and social development."

3.3 MAURICE ALBERTSON AND THE U.S. PEACE CORPS

More directly relevant than Hoover or Carter, however, was the life and work of civil engineer Maurice (Maury) Albertson (1918-2009). Growing up during the Great Depression in Iowa, Albertson was strongly influence by a family commitment to try to live out the Christian message of the Sermon on the Mount and by witnessing the impact of an extended drought on farmers and their communities. This led him to study water resource engineering and earn a doctorate from the University of Iowa.

After graduation, in 1947 he joined the faculty at Colorado State University, where he helped found the department of civil engineering and develop its focus on hydrology.

Maurice L. Albertson
Photo courtesy of Colorado State University

Figure 3.2: Maury Albertson (1918–2009), an engineer contributor to the founding the U.S. Peace Corps and a model humanitarian engineer.

Having been impressed with the way the Marshall Plan helped Europe recover after World War II, Albertson wondered, as he said in a 2008 interview, "Why not [something similar] with the rest of the world?" (Albritton, J., 2008). As a result, in the late 1950s he became a consultant to new U.S. Government efforts to promote development in Asia and was seconded to Bangkok, Thailand, where he helped establish a hydrological engineering program at what became the Asian Institute of Technology. Then in 1960 he was awarded a contract by the U.S. State Department to examine the feasibility of creating what was called a "point-four youth corps." The "point-four" referenced President Truman's fourth point in his 1949 inaugural address, which called for the United States to "embark on a bold new program" to "make available to peace-loving peoples the benefits of our store of technical knowledge in order to help them realize their aspirations for a better life" (as discussed in

Chapter 1) — a commitment that Democratic Senator Hubert Humphrey had promoted with the vision of a volunteer youth corp to provide technical assistance in developing countries. Albertson's co-authored report, expanded into book form, became *New Frontiers for American Youth: Perspective on the Peace Corps* (Albertson et al., 1961). The book explicitly describes the Peace Corps as extending the reach of volunteer Christian international service organizations into the promotion of American political ideals and lists among its Principal Project Needs, "engineering (irrigation, community water supply, flood control, roads, surveying, bridges)"(Albertson et al., 1961, p. 39).

As an upshot of his report, Albertson was asked by R. Sargent Shriver, the first director of the Peace Corps, to head a panel that would lay out many of the operational structures which, in short order, had over 10,000 volunteers serving in some 50 countries. Albertson subsequently became a consultant to such agencies as the World Bank, the United Nations Development Program (UNDP), and the United Nations Educational, Scientific and Cultural Organization (UNESCO) with a persistent focus on water and sanitation, farm water management and village development, and appropriate technology. In all cases, Albertson emphasized, as he said when awarded an honorary degree in 2006,

> We need to be motivated by service as well as by profit. We serve best by finding out what people want and helping them work to realize their dreams, not by going into a country and telling villagers what they need (Press Release, 2006).

In the words of Sandra Woods, Dean of Engineering at Colorado State, Albertson "was a world leader in water research [who continued] to work for what he [believed was] right and for the benefit of humankind" (Press Release, 2006).

3.4 *MÉDECINS SANS FRONTIERS* AND ENGINEERS WITHOUT BORDERS

Perhaps even more influential than individuals, however, has been the model of the NGO known as of *Médecins sans Frontiers* (MSF or Doctors without Borders). Hoover, Carter, and Albertson all fundamentally accepted, even when they were frustrated by, the notion of national sovereignty. The U.S. Peace Corps, with which Albertson was so much involved, is actually an agency of a sovereign country and thus tends to reinforce the whole concept of sovereignty or the idea that national governments have the final say over what goes on within their state boundaries. At the same time, from its beginnings, humanitarianism involved a questioning of the idea of sovereignty and associated ideas such as national patriotism and sacrifice. One of the fundamental tenants of MSF, which was founded in 1971, was to criticize and reject the primacy of national sovereignty as a final arbiter of boundaries for humanitarian action. MSF activists are committed to going where the problems are, even without the permissions of national governments, and to exposing the misbehaviors of governments toward their own peoples, insofar as these involve mistreating their citizens or depriving them of protection and care.

Stimulated by the ideals of MSF, the late 20th century also witnessed emergence of a host of other MSF-like NGOs: *Aviation san Frontiers* (1980), providing air deployment for humanitarian projects, Pharmacists without Borders (1985), Reporters without Borders (1985), Education without Borders (1988), Translators without Borders (1993), Lawyers without Borders (2000), Sociologists without Borders (2001), Chemists without Borders (2004), MBAs without Borders (2004), Librarians without Borders (2005), Farmers without Borders (2007), Scientists without Borders (2008), and Astronomers without Borders (2009). Yet one of the strongest parallel without-borders organizational developments has been associated with some form of the name "Engineers without Borders," in which engineering students and their professors began independently to explore possibilities of humanitarian engineering in diverse localities: *Ingénieurs sans Frontiers* (France, 1982), *Ingénieurs Assistance Internationale* (Belgium, c.1987), *Ingeniería sin Fronteras* (Spain, 1990), *Ingenierer unden Graenser* (Denmark, c.1992), *Ingenjörer och Naturvetare utan Gräser-Sverige* (Sweden, c.1995), Engineers without Borders (UK, 2001), Engineers without Borders (Australia, 2003), *Ingenieure ohne Grenzen* (Germany, 2003), *Ingegnería senza Frontiere* (Italy, c.2005), and others. In 2003 a number of these groups organized "Engineers without Borders — International" as a network to promote "humanitarian engineering ... for a better world," now constituted by more than 41 national member organizations.

Complementing such interests among engineers, humanitarians have increasingly come to see engineering and technology as having increasingly crucial roles to play in the world of humanitarian action. As one of three authors in a collection of studies reflecting on *Technology for Humanitarian Action* have put it, technologies that have been used for war must be developed for peace.

> For the future, there is a lot of potential for adapting and creating technologies for humanitarian ends, but new technologies will not automatically be put to humane uses without the political will and the economic means to do so. This necessitates building upon and furthering the ... trend of enlargement of humanitarian concern and expanded organizational effort [since the middle of the 20th century]. It means mobilization of the new culture to encourage the wealthy part of the globe ... to make the economic sacrifices necessary to create and apply technology in effective ways (Cahill, K., 2005, p. 19).

3.5 HUMANITARIAN ENGINEERING: CORE FEATURES

As Chapter 2 argued, humanitarianism has gone through a number of developmental phases. It is the more recent phases, from the latter decades of the 20th century, that have constituted a new context for the practice of engineering. From the Cuny, Albertson, and MSF stories of this period, one can abstract some key attributes of the humanitarian engineering ideal that pick up and emphasize especially the notions not just of crisis intervention humanitarianism but also vulnerability reduction leading to more rapid crisis recovery and even crisis prevention.

The central feature of the humanitarian movement as a whole has been the exercise of active compassion for those on the margins of social wealth and power. This marginality can be temporary or more long-term, but in either case humanitarian action aims to serve the well-being of otherwise under-served populations.

Engineering itself has been described as design within a context or under constraints — constraints largely imposed by physical, political, cultural, ethical, legal, environmental, and economic phenomena. Insofar as this is the case, humanitarian engineering may conveniently be described as working to escape what has been called the "social captivity of engineering" by capitalism or nationalism or some other form of wealth and power (Goldman , 1991; see also Johnston et al., 1996). In doing so, however, humanitarian engineering seeks to work within a new self-imposed constraint of seeking to help meet the basic needs of under-served populations. In brief, humanitarian engineering in the most general terms is the artful drawing on science to direct the resources of nature with active compassion to meet the basic needs of all — especially the powerless, poor, or otherwise marginalized.

CHAPTER 4

Humanitarian Engineering Education

A new form of engineering education is needed, one that covers a wide range of technical and non-technical issues, including water provisioning and purification, sanitation, public health, power production, shelter, site planning, infrastructure, food production and distribution, and communication. ... The challenge of creating a sustainable world demands a new and holistic look at the role of engineering in society ... to allow all humans to enjoy a quality of life where basic needs of water, sanitation, nutrition, health, safety, and meaningful work are fulfilled.

— Bernard Amadei and William A. Wallace, "Engineering for Human Development" (2009), pp. 7 and 10.

The development of humanitarian engineering education is a natural follow on to the rise of interest in humanitarian engineering. Such education will obviously benefit from an appreciation of engineering as a context dependent, externally constrained activity, as well as from some general knowledge of the history and development of humanitarianism. The material in Chapters 1 and 2 can thus be expected to function as useful components of any humanitarian engineering educational activity.

Examples of humanitarian engineers and their achievements will naturally contribute further to humanitarian engineering education. Some of the material in Chapter 3 can therefore be used and supplemented with other biographical profiles in ways that advance appreciation of humanitarian engineering as the artful drawing on science to direct the resources of nature with active compassion to meet the basic needs of all — especially the powerless, poor, or otherwise marginalized.

However, if chapters on engineering, humanitarianism, and the ideal of humanitarian engineering have had to simplify the complex subject matters with which they dealt — which they did — even more simplification will be required in our discussion of humanitarian engineering education. Education takes place from a diversity of perspectives and within multiple contexts that can seldom be captured in a syllabus. Nevertheless, engineering is itself a practice that depends on and regularly makes use of simplification and abstraction in the form of quantitative and qualitative modeling; engineers readily appreciate the simultaneous need for and limits of simplifying models.

The present chapter thus calls attention to various models of humanitarian engineering education and, drawing on our own experience, suggests a few other components that might well play useful roles in the design and implementation of humanitarian engineering education programs. But we would emphasize, as is the case with most engineering problems, that there is seldom one right way to design a humanitarian engineering curriculum. Instead, even more so in this regard than in many others, there is a recurring need to take clients, aspirations, resources, and context into account.

4.1 A FEW MODEL PROGRAMS

Given the more or less spontaneous emergence of a spectrum of humanitarian engineering interests and initiatives, engineering education could be expected to attempt a diversity of engagements. In many cases EWB activities, which are primarily project based, have become associated with educational programs, especially service-learning activities. A selective sampling of illustrative cases in diverse institutional settings includes:

- Technology Assist by Students (TABS), founded in 2000 at Stanford University;
- Engineering World Health (EWH), founded in 2001 at Vanderbuilt University and currently based at Duke University;
- Engineers for a Sustainable World (ESW), founded 2001 at Cornell University, now headquartered in Oakland, CA; and
- Engineers in Technical and Humanitarian Opportunities of Service (ETHOS), created in 2004 at Iowa State University.
- Humanitarian Engineering Leadership Projects (HELP) is a "student-run organization at Dartmouth's Thayer School of Engineering, founded in response to a simultaneous, growing need for service and engineering opportunities and the growing demand for global poverty reduction through local and sustainable solutions."

Sustainable Development Engineering

Sustainable development is a concept promoted by the World Commission on Environment and Development (1987) to bridge the divide between environmental critics of economic development and defenders of development especially in developing countries. The demand for some bridging concept became obvious at the UN Conference on Human Development (Stockholm, 1972). There proponents of *Limits to Growth* (Meadows et al., 1972) arguments based in criticisms of overpopulation, over exploitation, and over pollution were challenged by pro-growth arguments that appealed to both social justice and technological optimism.

A UN Commission created to deal with this challenge, chaired by Gro Harlem Brundtland (and thus sometimes called the Brundtland Commission), issued its report, *Our Common Future* (1987). This influential report shifted the environmentalist argument about what *should not* be done (development that causes environmental degradation) to what *should* be done: "sustainable development," defined as development that "meets the needs of the present without compromising the ability of future generations to meet their own needs" (p. 8).

Efforts to shift the discussion in this direction had occurred previously in a *World Conservation Strategy* (1980) formulated by the International Union for Conservation of Nature and Natural Resources. It was then carried forward in *Agenda 21* (meaning agenda for the 21st century), a product of the Earth Summit held in Rio de Janeiro in 1992. Indeed, by the mid-1990s a large number of heterogeneous groups had laid claim to being able to chart a path to sustainable development — groups that ranged from proponents of alternative technology and organic agriculture to free market capitalists and neoliberal economists. Sustainable development had undergone such conceptual inflation so as to became what Uwe Pörksen (1995) termed a "plastic word." It could mean almost anything.

Attempts to avoid this slide into vacuousness led to hard questioning from both the left and the right. From the left came the question, "Sustainable for what?" Simple material sustainability must be subordinate to some larger sense of the good. From the right emerged claims that sustainability was not enough. There is no sustainability without growth and innovation.

Sustainable development can nevertheless be defended as providing a framework — like other concepts such as goodness and justice, which can also be interpreted in many different ways — for dialogue over how engineering is to be pursued and practiced (Mitcham 1995). Sustainable engineering, for instance, might well be described as engineering that tries to take account of both environmental constraints and human aspirations for enhanced material welfare especially among the disadvantaged.

Figure 4.1: Cover of the Brundtland Report, which contains what has become the most widely adopted definition of sustainable development.

More closely linked with curriculum developments have been

- Engineering Projects in Community Service (EPICS), created in 1995 at Purdue University, as an undergraduate program to "create partnerships between teams of undergraduate students and local community not-for-profit organizations to solve engineering-based problems in the community," which was recognized in 2005 with the Gordon Prize at the National Academy of Engineering; and
- Engineering for Developing Communities (EDC), founded 2001 at the University of Colorado, Boulder, as a graduate program to educate "globally responsible graduate engineering students and professionals who can offer sustainable and appropriate solutions to the endemic problems faced by developing communities worldwide."
- Humanitarian Engineering and Social Entrepreneurship (HESE), established at Pennsylvania State University in 2006 to redress the imbalance of "the fact that 90% of the engineering that is done in the world today is directed primarily towards 10% of the world's population." The Penn State HESE program is also the institutional home of the electronic *International Journal for Service Learning in Engineering* (2006-present).

There are other illustrations from outside the United States. In Japan, for instance, the Tokyo Institute of Technology offers undergraduate and graduate programs in International Development Engineering to help "students become engineers who have ability, courage, and leadership, and can solve the problems" associated with international development projects. In developing countries there are cases where it can be argued that whole departments or schools constitute humanitarian engineering programs.

4.2 THE PEACE CORPS MASTER'S INTERNATIONAL PROGRAM

In light of the prominence accorded to it in Chapter 3, an engineering education program developed in conjunction with the Peace Corps deserves special mention. Despite some left-leaning criticisms of its association with the foreign policy agenda of the United States, the Peace Corps clearly exhibits a humanitarian dimension and has learned from and incorporated the experiences of related NGOs such as the American Friends Service Committee (AFSC, founded in the US in 1917) and Voluntary Service Overseas (VSO, founded in the UK in 1958, but eventually becoming international). Drawing on its own and related experiences, in the late-1980s the Peace Corps began working with universities to create a graduate Peace Corps Master's International (PCMI) degree program combining a year of campus residency followed by two years of Peace Corps service in the field. Although the program does not use the term "humanitarian engineering," it nevertheless offers a useful perspective on related educational efforts.

Although there are now almost 100 participating academic units at more than 40 institutions such as Colorado State University, where Maurice Albertson (a Peace Corps founder discussed in

Chapter 3) taught for many years, University of California Davis, and George Mason University, the PCMI program at Michigan Tech can be taken as a leading, representative example.

Michigan Tech initiated its PCMI program in the mid-1990s and as of 2009 offers seven PCMI programs in Applied Natural Resource Economics, Civil and Environmental Engineering, Forest Resources and Environmental Science, Mechanical Engineering, Natural Hazards Mitigation (Geology), Rhetoric and Technical Communication, and Science Education. As the program flier notes, "Contributing to development requires more than technical skills, it requires dedicated individuals serving as the human face of development — working on a daily basis with communities around the world."

Students apply simultaneously to the Peace Corps and to one of the seven PCMI Michigan Tech programs. Once accepted into both the Peace Corps and a PCMI program, they spend a year at Michigan Tech beginning work on the required 30 semester credits, two of which are dedicated to research on the Peace Corps assignment with at least another seven coming from work performed during the two-year Peace Corps field assignment. Along with technical engineering course work, students must also complete a course in Rural Community Development Planning and Analysis and some work in a relevant foreign language. Additionally, students are required to take at least one "cultural sensitivity elective" from a list that includes such titles as "Communicating across Cultures," "Developing Societies," and "Anthropology of Science and Technology."

The last title is particular appropriate, because in his Peace Corps book, Albertson et al. (1961) explicitly quoted from a volume by anthropologist Edward H. Spicer on *Human Problems in Technological Change* (Spicer, E., 1952), a classic collection of case studies examining difficulties and unintended consequences of technological development. As Albertson noted, the importance of a cultural anthropological perspective in "Peace Corps orientation and training cannot be over-emphasized" (Albertson et al., 1961, p. 81) It is clear that in the Michigan Tech program as well, it is understood that technical knowledge must not be allowed to crowd out social science and the humanities in the PCMI curriculum. As has been repeatedly observed in humanitarian engineering innovation, cultural issues are at least as important as technical ones. At Michigan Tech non-technical study includes as well the ethical guidelines for human subjects research, because many students will include this as an element of their field research, and readings in human organization and management.

Finally, it can be noted that PCMI students have the option of earning a Graduate Certificate in Sustainability. Issues of sustainability cannot help but play a major focus in almost any humanitarian engineering work. Along with their course work, students also meet regularly with residents in a Peace Corps Fellows program and other returned Peace Corps volunteers to discuss issues related to their forthcoming assignments. Once they have completed their two years of field assignment — the period in which they undertake what can be termed a humanitarian engineering project — PCMI students return to Michigan Tech to complete a thesis and be awarded their degrees.

Still another humanitarian engineering spinoff from Peace Corps service is the work of Amy Smith, a senior lecturer in Mechanical Engineering at the Massachusetts Institute of Technology.

After earning an MIT BS in Mechanical Engineering in 1984, doing four years of Peace Corps service in Botswana, and returning to MIT to earn her MS degree, Smith founded the D-Lab (Development through Dialogue, Design, and Dissemination) program to promote appropriate technology for developing countries. She also runs MIT's IDEAS competition in which teams of student engineers "design projects to make life easier in the developing world." For her work, Smith has been profiled in the *New York Times* (Kennedy, P., 2003) and *Wired* (Dean, K., 2004) and recognized with a MacArthur Foundation "genius award" (2004), clearly providing another biographical model for humanitarian engineering education.

4.3 WHAT COUNTS AS A HUMANITARIAN ENGINEERING PROJECT

Deciding what truly counts as a humanitarian engineering project is not always easy. Efforts to clarify understandings in this regard within the CSM undergraduate Humanitarian Engineering Minor program have led to the formulation of a set of four guiding criteria:

- One, there must be a need that originates with the people directly benefitting from any proposed work.
- Two, whatever need is involved should be related to a basic human need, although it is also possible to include higher level needs such as education and economic development.
- Three, good communication is essential, preferably with the people directly benefitting from the work and/or commonly through an NGO intimately familiar with the local context.
- Four, the need should be one that can benefit from engineering skill and knowledge.

One way to operationalize the first criterion is to use the engineering design process systematized as the "quality function deployment technique" (see Cohen, L., 1995). The fundamental idea is to begin by identifying stakeholders and then working with them to establish a set of prioritized needs. Subsequent analysis compares competing solutions and finally, based on such inputs, design specifications are developed.

The second criterion is more problematic than it may initially appear to engineers. The reason is that human needs depend on human interpretations, which in turn are strongly influenced by cultural beliefs about the nature and meaning of human life. Nevertheless, from a perspective that necessarily reflects American engineering beliefs, there exists a hierarchy of physiological needs, defined by survival time for anyone denied access to a number of basic life requirements. Humans can live only a few minutes without air, so this represents the most pressing physiological need, followed closely by warmth or thermal regulation. People living on the extreme latitudes of the planet would be concerned with thermal regulation during the winter months. However, thermal regulation can also be a problem associated with over heating in tropical or desert regions. Need for water and food would follow, respectively, as graphically represented in the model of Figure 4.2.

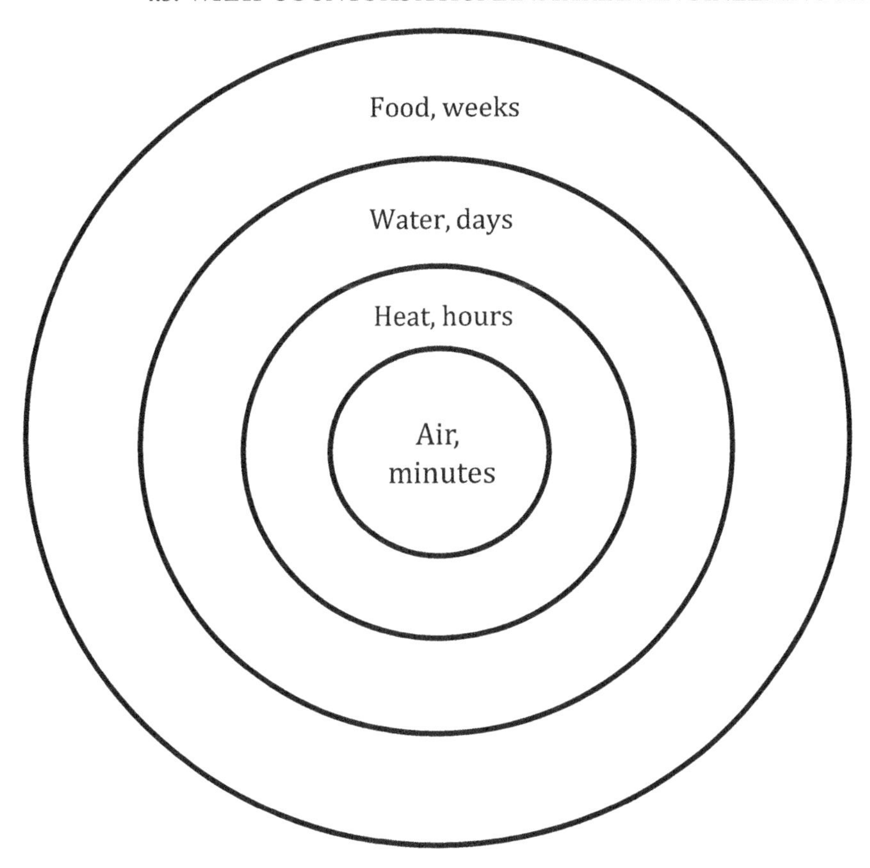

Figure 4.2: A nested hierarchy of physiological needs.

From an engineering perspective, it is easy to argue that inner circle needs are more basic than others. An emergency response or triage worker must keep such a hierarchy in mind when responding effectively to disasters (Davis and Lambert, 2002). At the same time, to think only in these terms would significantly limit humanitarian engineering. It is thus necessary to move beyond such immediately physiological or technical considerations, to psychological, social, and political concerns, when thinking about basic needs.

Working with criteria three and four promotes, even more than criterion two, appreciation of the degree to which psychological, social, and political aspects of a project are often as much if not more crucial than technical ones. Communication is crucial among all those involved in humanitarian engineering projects, engineers and non-engineers alike. So education in communication skills that go beyond abilities in simple technical communication are important. Communication has to be oriented not just toward the facilitating of technical team effectiveness but toward the creation of interdisciplinary and interpersonal community. Such a recognition promotes deeper understandings of (sustainable community) development (Bridger and Luloff, 1999). If projects really are to benefit others, it is crucial to seek out local sources of knowledge and to value them, which can sometimes demote the importance of technical engineering skills and knowledge. This idea, known as participatory action research, is an extension of ideas from Freire, P. (1970), and has been elaborated by Stephen Biggs (see sidebar; see also the analyses in Fals-Borda and Rahman, eds., 1991). In order for any humanitarian engineering project to be socially sustained, there must be ownership on the part of the local people. A major source of ownership comes from the engagement or participation of the local people in all aspects of the design process.

A Spectrum of Participatory Research

(adapted from Stephen Biggs, 1989)

Participatory research, which can include engineering design and construction work, includes a spectrum of at least four modes of participation:

1. *Contractual*: People are contracted into the projects of the researchers to take part in their inquires or experiments.
2. *Consultative*: People are asked their opinions and consulted by researchers before interventions are made.
3. *Collaborative*: Researchers and local people work together on projects designed, initiated, and managed by researchers.
4. *Collegial*: Researchers and local people work together as colleagues with different skills to offer, in a process of mutual learning where local people have control over the process.

As a result, it is possible to conceptualize a need among engineers to look for opportunities to help build capacity for autonomous action among those with whom they work. In this regard, a series of questions adapted and expanded from Baillie, C. (2006) can serve as a template for self-examination. In thinking about any project, it is useful to ask:

- Who benefits and who pays?
- Who stands to gain or lose?
- Who says who needs what and when?
- Who is contributing to the design and implementation?
- How will the project be sustained?

A closely related set of questions for bridging engineering ethics (as referenced in Chapter 1) and humanitarianism (as described in Chapter 2) has been formulated as "Questions for Humanitarian Engineering Ethics Education" (see sidebar).

Questions for Humanitarian Engineering Ethics Education Projects

(with explanatory justifications)

1. Does this engineering design work promote the good of all people independent of their nationality, religion, class, age, or sex?

   Justification: Humanitarianism as an ethical tradition historically rejects the significance of such distinctions.

2. How might this engineering project be related to the protection and promotion of human rights?

   Justification: Humanitarianism has been repeatedly linked with the emergence of human rights especially as recognized in such documents as the Universal Declaration of Human Rights (1948).

3. Is the product, process, or system being engineered likely to help meet a humanitarian crises such as those typically associated with war or natural disasters?

   Justification: Humanitarianism is often exemplified with humanitarian aid during such crises.

4. Is this engineering design work addressed especially to meet some fundamental human needs (understood as those for water, food, and shelter)?

   Justification: Humanitarianism regularly argues the priority of fundamental needs over needs associated with affluence.

5. Is this engineering design work oriented toward providing benefits for those otherwise underserved by engineering either in the advanced or the developing regions of the world?

   Justification: Humanitarianism typically manifests what is known as the "preferential option for the poor."

6. In what ways might the engineering design and construction work be more compatible with not-for-profit enterprises than for profit enterprises? How might such engineering and construction work that did seem more compatible with the pursuit of economic profit be either supported by alternative means or recast so as to be compatible with economic motives?

   Justification: Humanitarianism has often been practiced in tension with corporate economic interests.

7. What is the likelihood that this engineering product, process, or system will be sustainable?

   Justification: Humanitarianism is often thought to be supportive of and appropriately pursued in synthesis with sustainable development.

Qualifications: Not all question will be equally relevant or significant in all courses. Instructors in any one course might well select from this set. Additionally, when asking such engineering design questions, instructors should help students reflect on and critically examine the arguments and justifications referenced here. There should be no uncritical acceptance of any arguments and justifications.

4.4 THE NEEDS QUESTION

Returning to the working definition of humanitarian engineering as the artful, active compassionate utilization of resources in nature to meet the basic needs of all, especially the marginalized, raises the question of needs in an even more complex way than mentioned with regard to criterion two above. Three elements are involved in this definition: the motivation of active compassion, the focus on basic needs, and the concern for all persons. Although the central element is basic needs, which the last element ties to other persons, the first element also implicitly references needs: a need in humanitarian engineers themselves. They need to act with compassion. Do they also have a "basic need" to do so? "Need" can become a concept that bridges humanitarian engineers and those they aspire to serve. Especially is this so insofar as it is appreciated that what counts as need or motivation depends on the act of human interpretation, which again is strongly influenced by cultural beliefs. Americans, especially American engineers, typically think that some things are more important or needful than may be the case with peoples in other countries and cultures. For instance, it is not clear that all peoples have a need for economic development; for most of human history, people failed to express such a need.

One approach to the study of needs that can begin to illuminate the complexity of needs thinking is associated with the life and work of the humanist psychologist Abraham Maslow (1908-1970). The eldest child of a Russian-Jewish immigrant family in Brooklyn, New York, Maslow studied with a number of leading psychologists, first at the University of Wisconsin, then at Columbia University, and became the founder of a "third force" (after psychoanalysis and behaviorism) among American psychologists. His emphasis on the importance of wellness over illness (contra psychoanalysis) and intentional action over stimulus-response (contra behaviorism) was manifested in research into efforts by human beings to achieve their highest potential. His research resulted in such concepts as the hierarchy of needs, meta-needs, self-actualization, and peak experiences. For present purposes what is most useful is his analysis of a human needs hierarchy.

One feature of Maslow's work was an effort to discover what it means to be a fulfilled human being, focusing not on the mentally ill but what he considered exemplary individuals — including people often described as humanitarians, such as Jane Addams, Albert Einstein, and Eleanor Roosevelt — Malsow put forth a model of human motivation that relied neither on sex and sublimation (as in Freud) nor physiological training (as in Ivan Pavlov and B.F. Skinner). Instead, he developed a hierarchy that moved from physiological needs for air, water, food, and shelter (as mentioned above), through needs for safety, social interaction, and esteem, to notions of self-actualization and self-transcendence. Figure 4.3 summarizes this hierarchy in the form of a pyramid.

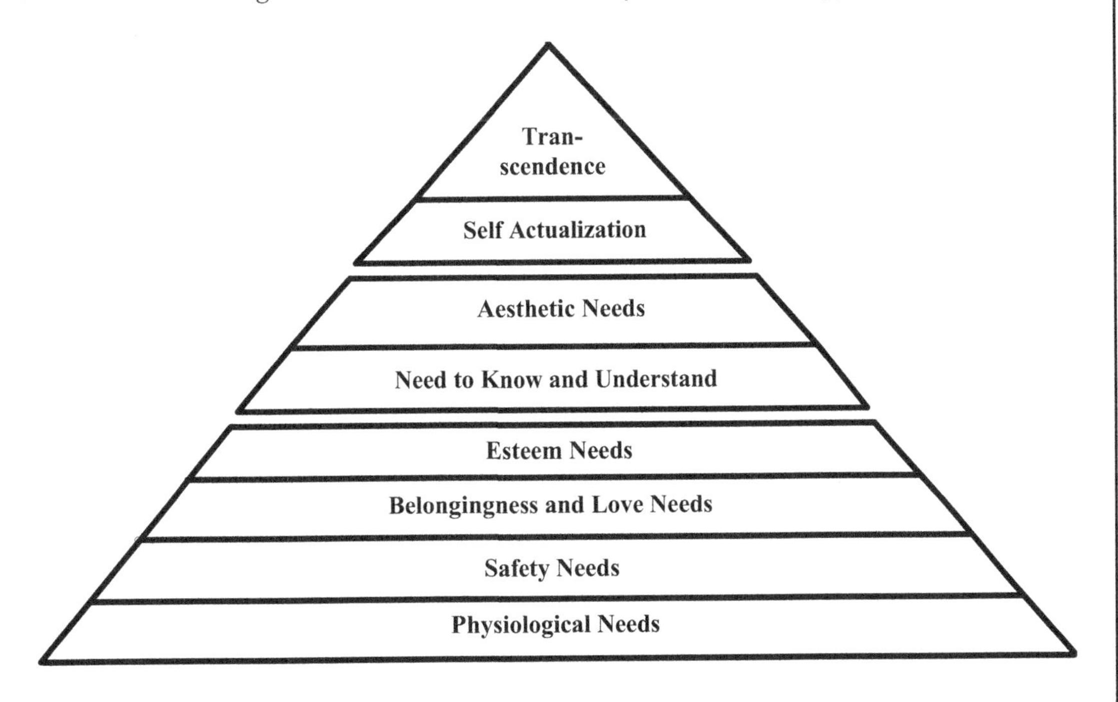

Figure 4.3: The hierarchy of human needs (adapted from Maslow, A., 1954).

Simplified, Maslow can be interpreted as arguing that the needs at the bottom of the pyramid (sometimes called deficiency needs) more often function as stimuli to behavior while those at the top (sometimes called abundance needs) can serve as motivations for actions, and that as people satisfy lower level needs they are free to pursue higher-level needs. That is, as individuals work to satisfy deficiencies real, potential, or imagined with regard to thirst, hunger, physical safety, and social group membership, they become progressively liberated to pursue and experience more profound forms of abundance in knowledge, aesthetic appreciation, and related forms of self-actualization. Indeed, Maslow generally argued for a progression within humans, from the lower levels of the pyramid to the higher. Relationships can also readily be drawn between Maslow's third force human potential psychology and the positive psychology research program on human happiness (see, e.g., Kahneman et al., 1999). (The Maslow approach to conceptualizing needs has, however, been challenged; see sidebar.)

Needs, Wants, Desires, Interests

The concept of needs is more complex and contentious — and not always as objective — as it initially appears. For instance, although human beings may have an objective, universal need for food, what counts as food can be interpreted quite differently by different human beings in different societies and different historical contexts. A hungry vegetarian might well starve rather than eat meat; a devout Muslim might go hungry rather than share a meal of barbeque pork and beer. Efforts to use the concept of needs as an objective guide for human development programs have thus come up against challenges from those who claim that developed countries often project their interpretations of appropriate need satisfaction onto others.

One effort to reflect on the problematic character of needs language concludes that an important

> question for needs methodology is the relationship between human beings and their needs. The underlying problem is how to decide if the acknowledgment of needs represents only one part of the human being and, if so, which part. This requires a careful analysis of the contradictory influences on needs perception and their integration into the whole of human existence and fulfillment (Galtung and Lederer, 1980, p. 345).

One contribution toward meeting this challenge would be to distinguish needs, wants, desires, and interests. The concept of needs can be interpreted in objective or historical terms, with the putative objective sense often obscuring a historical one. By contrast, the concepts of wants and desire seem more malleable and readily connote historicity and subjectivity without qualification, but in ways that may also stigmatize subjectivity. Wants can be manipulated by advertizing. Desires are easily criticized precisely because they seem so subjective and are often attached to lower level behaviors, as when people are said to have a desire (but not a need) for physical pleasure. In extreme cases desires even become addictions, as in the case of alcoholism.

The concept of interests, however, admits to historicity and subjectivity while not being so easily attached to physical pleasure. Indeed, human interests imply some degree of objectivity and historicity without (like need) obscuring the latter with the former or (like desire) making everything subjective. It is precisely for this reason that in philosophical anthropology — as manifested in pragmatism, existentialism, and critical social theory — notions of human interest play a more prominent role than needs. Jürgen Habermas (1968), for instance, argues for recognition of three basic human interests: technical, practical, and emancipatory. It would be useful to explore in what ways engineering, which Habermas identifies primarily with technical interests, could also be related to emancipatory political interests.

Linking needs, wants, desires, and interests is the concept of quality of life. This concept too has been shown to have considerable cultural variability (Hofstede, G., 1984).

Given this model, it is tempting to think of the motivation of humanitarian engineers is situated on higher levels of the hierarchy, with the aim of meeting the lower level needs of those being served. This is one possible interpretation. At the same time, there is something invidious if not insulting in a framework that winds up placing those on the initiating side of humanitarian work on a higher psychological level than those on the receiving side. Moreover, the very hierarchy itself — and especially its imputed dynamics — surely reflects to a considerable extent the beliefs and assumptions of American engineers operating in the context of what has often been described as a needs-based materialistic culture (see, e.g., Illich, I., 1978). What, we may ask, are the relations between typically modern discussions of need in contrast with more traditional notions of the good? Maslow's model may thus function not only as an explanatory model but also as a framework for self-questioning.

One result of such introspection might be a critical reformulation of ideas about how we live in the "developed world" and the imagination of models for environmentally sustainable living in a sustainable global economy. After considering the energy and material intensive aspects of our own society, one cannot help but question its use as a model — and even the terminology "developed" versus "developing." Surely we are all developing as we seek improved community models for global sustainability. Additionally as one aspiring humanitarian engineer noted with regard to himself:

Initially, I was excited about [humanitarian engineering] because of the opportunities to design appropriate technologies for needy international communities. While this excitement does still exist, [after study and experience] I am much more leery; during the process I learned a lot about technology in society, the need to challenge structures, the need to work in one's own community, and the dangers of international placements (VanderSteen, J., 2008, p. 288).

4.5 NEW DIMENSIONS IN ENGINEERING AND EDUCATION

Although subject to continuing debate, the basic dimensions of humanitarian engineering may thus be summarized as follows. While advances in science and technology have benefitted many persons, even the World Bank (2008) admits that they have done little to decrease rich/poor divides, a situation to which a number of specific organizations have tried to respond. Among these, many emphasize the use of engineering expertise. Humanitarian engineering projects, typically operated on a not-for-profit basis, aim either to provide basic needs such as food, water, shelter, and clothing, when these are missing or inadequate in the developing world, or higher-level needs for under-served communities in the developed world.

In contrast to corporations, which aim for relatively near-term profit, and governments, which fund in light of election cycles and constituent dependencies, humanitarian projects will typically be of longer-term importance for society as a whole. Humanitarian science and engineering ideally engage local communities in direct participation in determining project needs and directions, thinking in terms not of years but of decades. Additionally, they seek strategies, designs, and technologies that promote both the sustainability of natural systems and cultural traditions (see, e.g., Azapagic et al., 2004 and Mulder, K., 2006).

As Bernard Amadei and William Wallace have argued, there are needs for new forms of engineering education, what we have called humanitarian engineering education, to prepare students to meet such challenges. This may well constitute a transformation not simply in engineering education but in education more generally. It is also a need that deserves some place on any hierarchy of needs. Interestingly enough, its placement could be argued to belong both on the lower levels — insofar as humanitarian engineering will at least help people meet needs for water and shelter — and on higher ones — insofar as engineers and others undertake the creation of new curricula on the basis of a need to know and understand the realities of the larger global world in which they live, a world that includes a greater percentage of less powerful and wealthy than has often been recognized.

CHAPTER 5

Challenges

For many [humanitarianism] is best identified with the provision of relief to victims of human-made and natural disasters. For others, though, humanitarianism does not end with the termination of the emergency. ... No longer satisfied with saving individuals today only to have them be in jeopardy tomorrow ... many organizations now aspire to transform the structural conditions that endanger populations. Their work includes development, democracy promotion, establishing the rule of law and respect for human rights, and post conflict peacebuilding. These more ambitious projects expand [humanitarian work] — but, for better and worse, they coincide with and sometimes become part of the grand strategies of many powerful states.

— Barnet and Weiss, eds., 2008, *Humanitarianism in Question*, p. 3.

Humanitarianism has become a progressively expansive project. As part of its expansion, it now includes engineering. But like any expansion, this one brings with it not only benefits but also challenges.

Engineers have long been focused on meeting the needs of humanity. During the past century, however, we have become increasingly aware of new dimensions of human need not regularly addressed by engineering practice and education in Europe and North America. Though we have been to the Moon, sent robots to Mars, and can do marvelous things on the mega and nano scales, the fact remains that on our special planet, roughly one fifth of the people lack access to clean water and 11 million children under the age of five die every year from malnutrition and disease (Kandachar and Halme, 2008). Indeed, even the engineered infrastructure of the developed world will need to change in order to secure a sustainable future. Our energy and water consumption, food production, and patterns of material consumption will need to change to accommodate a habitable world for our descendants. Engineering graduates must understand the global constraints they face — physical, economic, environmental, social, political, and cultural, all together — and how to use all available tools as they work toward socially just and sustainable solutions. This is a challenge to which humanitarian engineering can make a meaningful contribution. In pursuing the goals of humanitarian engineering and humanitarian engineering education, it is nevertheless useful to consider two kinds of challenges: practical and theoretical.

5.1 PRACTICAL CHALLENGES

Practical challenges for any humanitarian engineering education program run the gamut from what to include in a curriculum and how to include it, to the difficulties of interdisciplinarity, travel, communication, and finances. There will be basic differences between undergraduate and graduate programs. Following are some necessarily brief comments on each of these areas, related mostly to undergraduate program development.

Figure 5.1: Two engineering students study a water distribution drawing prior to laying the pipe while local girls look on.

The first challenge in developing an undergraduate humanitarian engineering program is figuring out what to include and how to manage the inclusions. The engineering curriculum is already full — indeed, over full. It takes most students more than the standard four years to earn a bachelor's degree. Although a stand alone program designed from scratch might be ideal, since humanitarian engineering is not so much another kind of engineering as something to be infused into other engineering programs, the ideal is not a likely option and perhaps not even desirable. Besides, engineers are used to working under constraints. Designing a curriculum is a typical engineering-like problem.

Since undergraduate engineering programs include humanities and social sciences requirements and have at least some room for electives, one open path is to begin with these. The general agreement that historical, social, and cultural education is crucial means it will be important to work with interested humanities and social science faculty to develop courses in topics such as cultural anthropology, sociology and economics of development, histories of colonialism and post-colonialism, theories of ethics and social justice, and more. Engineering faculty can then work with technical electives and senior design to develop more technical course work.

But interdisciplinarity — that is, the collaboration between humanities, social sciences, and engineering faculty — is not always easy. The contemporary academic world is heavily disciplined, in many senses of the term. The challenges of working in interdisciplinary teams are significant and related strongly to developing a sense of trust between the individual team members. There is thus a need to spend time together (in an often time-short environment) discussing issues, working on proposals, and disseminating information, sharing perspectives and experiences, producing scholarly work. One of our own valuable interdisciplinary experiences was a course on "Engineering and Sustainable Community Development," with collaborations from half a dozen faculty and a cohort of two dozen students. Most of the faculty contributed to the course as an overload because it was such a good experience.

Travel to and practical engagement in humanitarian engineering project sites are common components of humanitarian engineering programs. Effective planning must precede especially any foreign travel with students. Planning should include checking in with the institutional staff responsible for international programs, who may be able to help students take advantage of foreign exchange opportunities. All faculty involved and most students should be trained in first aid. Students should consult a doctor or local health clinic to determine recommended inoculations and/or prophylaxis. Faculty should have information for multiple contacts in locations to be visited, with advanced arrangements made for mobile telephone and/or satellite telephone service. It is useful to develop a packing list for students to guide their trip preparation.

With students, faculty leaders must exercise and encourage a heightened sense of awareness of surroundings. Pay attention to what is happening around you. Stay out of areas prone to violence. Hire a chauffeur (preferably one vouched for by an in-country contacts) and a large enough vehicle to safely transport everyone to and from work locations. Require that students work in pairs or preferably groups of three (if something happens to one then the second can go for help while the third administers first aid). Handheld walkie-talkies are often useful for in-field communications, since groups will often need to divide tasks so as to get more done over short periods of time. Identify the nearest hospital or clinic. Communicate daily with in-country contacts because they will be aware of local dangers (poisonous insects, plants, etc.) or imminent political instabilities. Become aware of and obey all local laws. An excellent reference for what to do in the case of the unthinkable accident in which a student or faculty member is seriously injured or killed is by Ajango, D. (2005).

Communication skills are one of the biggest challenges, and these come in multiple forms. All students should be required to learn the basics about any country they plan to visit: government structure and current leaders, major geographical features, climate, history, religion, foods, sports, customs, and cultural taboos. Everyone must be willing to dress and act modestly. There will be even greater challenges in non-English speaking countries. At least some faculty and students should be fluent in the local language. When recruiting students for a project, seek those with appropriate language skills. Remember to consider international exchange students or interns visiting your campuses as possible team members.

As another aspect of communication, it is useful to maintain relationships with the same local people over years or decades in order to develop a sense of trust among all involved. Challenges in this regard can nevertheless arise from changes in political leadership, illnesses, or death. Such changes will often require the patient reestablishment of ties with new leadership. Yet changes can also lead to new opportunities, as leaders take on new positions, perhaps at higher levels of government. Such changes accentuate the need for developing relationships with leaders at multiple levels.

To cite one example from a CSM humanitarian engineering project in Colinas de Suiza, Honduras: In 2008 Lic. Cecilio Santos, the 29 year old Director of the Water and Sanitation Department within the Municipality of Villanueva, died suddenly of cerebral aneurism. He was central to a water project being developed to supply roughly 10,000 people. His death was a devastating blow for all involved, but the project was able to continue through relationships already established with his subordinates at the Water and Sanitation Department, the Patronato (local elected leadership) of Colinas de Suiza, and the mayor of Villanueva. Then in elections the next year, the mayor was voted out of office, so that new relationships had to once again be established. But then the former mayor accepted a cabinet position with the incoming president, offering a new set of opportunities for future projects at the national level. The need for patience, flexibility, and prudence can seldom be over emphasized.

Expenses associated with humanitarian engineering education projects include (a) faculty time to develop and offer new courses, (b) graduate and undergraduate student assistance, (c) equipment, which run a spectrum from educational to project supplies, and (d) travel. Although such expenses can be reduced by commitment and good planning on the part of all involved, it is difficult to manage without some external support. Faculty can apply for grants to support course development as well as volunteer their time. Work study students can be supported through financial aid budgets. Some equipment donations can be solicited. Students often have to raise money to cover their travel expenses.

To give some idea of the kind of expenses involved, consider again the Colinas de Suiza project. Retail costs of construction materials, equipment rental, and necessary stilled services exceeded one million dollars. The project was only possible through a combination of in-kind donations (plastic pipe and fittings) from the Plastic Pipe and Fitting Association (an international association of plastic pipe and fitting manufacturers), human labor (sweat equity) from the local people and from the Water and Sanitation Department, and NGO grants (in our case, Food for the Poor, a Catholic based NGO based in Ft. Lauderdale, FL, and CEPUDO, their Honduran affiliate located in San Pedro Sula). Local people contributed $100 per family for the water storage tank. (This represents a two week salary for the average worker in the village.) The Municipality of Villanueva agreed to drill the well and install a pump. Humanitarian engineering students provided engineering support through mapping, water distribution system design, and the acquisition of in-kind donations.

Since many of these practical challenges grow out of humanitarian engineering projects, it is important to consider whether such projects are necessary components of the engineering educational experience. In his study of humanitarian engineering education, Jonathan VanderSteen argues strongly that humanitarian engineering be included in all engineering curricula. However,

> The main focus of the curriculum should not be on "helping" others, but on awareness, not so much of global poverty, but of the forces that cause marginalization. In an intensely individualistic society, humanitarian engineering must focus on relationships and operate with community in mind (VanderSteen, J., 2008, p. 295).

Still, the opportunities for learning to communicate with people from a culture and society other than our own offer an exceptional potential for learning that will undoubtedly help educate young engineers more sensitive to trans-technical issues of the engineering design process.

Figure 5.2: Approximately 50 local people of Colinas de Suiza are working to lay pipe for their water distribution system.

5.2 THEORETICAL CHALLENGES

Theoretical challenges have more to do with thinking than with doing. Before or after or in the midst of it, how should we think about humanitarian engineering? How should we think about helping others? The whole of this small book has proposed one way to think, a way that sees humanitarian engineering as a cross-fertilization between the historical developments of engineering and of humanitarianism. But there are other, counter ways of thinking that must not be ignored and that pose distinct challenges. They challenge engineers and those of us who endorse the idea of humanitarian engineering to ask questions about the fundamental viability of the humanitarian engineering ideal. In many ways they are akin to some of the basic questions that are also being asked within the humanitarian movement as a whole (as referenced near the end of Chapter 2).

With regard specifically to humanitarian engineering, however, one question can be formulated quite simply as follows: What if the people who are offered humanitarian engineering assistance reject it — especially when those who offer the help are genuinely convinced it will benefit those in need? Potential rejections can take many forms, from ignoring proffered assistance to passive aggressive resistance and sophisticated intellectual criticism. Rejections can also come from many different persons within a community. Some forms of rejection are unconscious, others conscious. Humanitarians will often have to struggle to appreciate and assess these rejections in their manifold complexity. They must be sensitive to, but need not necessarily accept at face value, all criticism.

Consider the example of Ivan Illich (1926-2002). In 1968 Illich, a Catholic priest and theologian who had worked in Latin American for a decade, gave a talk to a conference of humanitarian aid workers in Mexico that concluded with the following words:

> Use your money, your status, and your education to travel to Latin America. Come to look, come to climb our mountains, to enjoy our flowers. Come to study. But do not come to help.

A transcript of this talk, subsequently titled "To Hell with Good Intentions," is widely available on the internet. It reiterates in even more biting terms a criticism of what Illich termed Christian

Figure 5.3: Ivan Illich (1926–2002), an insightful critic of ideology of development.

"dogooding" articulated in an article, "The Seamy Side of Charity," published the year before in the Jesuit weekly *America* (Illich, I., 1967).

But Illich was an intellectual outsider, a cosmopolitan polymath who subsequently criticized his own first person plural identification with Mexicans. The reference to "our flowers" reveals that the transcript had not received the careful editorial corrections that Illich gave to anything he himself published. We must not assume too quickly that Illich is an authentic spokesperson for the Mexican poor. Similar qualifications may legitimately be entertained with regard to any intellectual who claims to speak for those who are the recipients of humanitarian interventions. This qualification deserves to be entertained for those who praise as well as those who criticize.

The most profound theoretical challenge is thus to come to terms with our own interests and intentions — our praise and our criticism — and to consider carefully our understandings of what we are about, insofar as we engage in humanitarian engineering, its associated educational programs, and its criticism. In this regard the series of questions for self-examination formulated by Schneider et al. (2009) in their critique of what they term "engineering to help" may be useful. Humanitarian engineers would do well, they argue, to ask themselves a set of six questions:

(1) What are your motivations?

(2) What is the history and context of development projects in the area?

(3) Who benefits and who suffers from the project?

(4) Who is held accountable?

(5) What are the possible unintended consequences?

(6) Do we view communities as "less-than"?

At the same time, intellectuals such as Illich (1967) — we can testify on the basis of personal experience that Illich himself would likely agree — must undertake their own examinations of conscience. A sense of righteousness should no more be allowed to inhibit ignoring the good that others do than to rationalize failing to critically examine oneself.

CHAPTER 6

Conclusion: Humanizing Technology

> [H]ow naturally inhumanity combines with technology. … We have experienced technology in the service of the destructive side of human psychology. Something needs to be done about this fatal combination. The means for expressing cruelty and carrying out mass killing have been fully developed. It is too late to stop the technology. It is too the psychology that we should now turn.
>
> — Glover, J., 1999, *Humanity*, p. 414.

The argument presented in this small volume may be summarized as follows. First, engineering is a profession classically and appropriately defined as an artful drawing on science to direct the resources of nature for the use and the convenience of humans. As such, engineering involves systematic, context dependent design work, operating under constraints. Historical and social determinations of the meaning of "use and convenience" are key external constraints.

Second, since the emergence of the engineering profession, social history — particularly in Europe but increasingly on a global scale — has given rise to what is called the humanitarian movement. This movement constitutes in part a critical reassessment of some common assumptions about use and convenience. Humanitarianism may be described as an active compassion directed toward meeting the basic needs of all. The fundamental belief is the need for medical care to relieve suffering, along with such basic goods as water, food, and shelter should be available to everyone, irrespective of distinctions of class, nationality, religion, race, sex, or other determination. Yet since the relevant goods are already available to the wealthy and powerful, as a practical matter humanitarianism focuses on addressing the basic needs of the powerless, poor, or otherwise underserved persons at the national or international levels.

Third, engineering is appropriately influenced by the humanitarian ideal, thus giving rise to opportunities for new understandings and a historically distinct practice of this technical profession. Insofar as this is the case, something called humanitarian engineering may be described in general terms as the artful drawing on science to direct the resources of nature with active compassion to meet the basic needs of all persons irrespective of national or other distinctions — an artful, active compassion that, in effect, is directed toward the needs of the poor, powerless, or otherwise marginalized persons.

Observe especially in this regard how the standard engineering ethical commitment to the protection of public safety, health, and welfare (as described in Chapter 1) connects with the humanitarian ideal (as spelled out in Chapter 2). The fact is that significant groups of people in all nations and throughout the world are lacking in provisions related to the basic goods of safety, health, and welfare. This would seem to oblige engineers to approach with some skepticism any career path that led to engagement with nothing more than the design and manufacture of high-tech toys for an advanced consumerist society.

Fourth, an educational curriculum in support of the practice of humanitarian engineering is in fact emerging in multiple quarters. This is nevertheless a complex and on-going process. Studies in the history of engineering, the development of humanitarianism, and profiles of practitioner models were suggested as especially appropriate elements in any humanitarian engineering curriculum. After referencing a number of emerging programs, we also argued that central to any strong curriculum would be the study and appreciation of interactions between engineering or technology and culture. Approaches to such study and appreciation include but are not limited to learning the rudiments of participatory action research and critical reflection on the meaning of human needs. The needs question we found offered a special opportunity to develop understandings of those we see as peoples in need, as well as ourselves as need-seeing persons. The relation between needs and goods — between need and the good — offered another salient entry point for humanitarian engineering reflection.

Fifth, humanitarian engineering education was admitted to be fraught with numerous challenges. We reviewed a few of these as they are manifested at the levels of practice and theory. Practical challenges span issues of time and money — not to mention the difficulties of interdisciplinarity, the dangers of foreign travel, and the problematics of supplementary language acquisition. Theoretical challenges call for critical reflection on, along with self-critical examination in the light of, diverse criticisms of humanitarian efforts. How to negotiate such demands is not easy. But along with a template of questions that would be humanitarian engineers might use to interrogate themselves, we suggested that utilization of and reliance on efforts in the humanitarian movement to formulate principled guidelines and standards for practice offer a further way to negotiate this challenge.

As an effort to promote discussion of humanitarian engineering, our modest book aspires to be another effort to promote a new form of engineering and to help address some of its challenges — not so much at the practical level as at the level of conscious reflection. The assumption is that if we think and clarify the meaning of humanitarian engineering, it is more likely to become an important feature of the spectrum of engineering activities that increasingly constitute the high-tech, advanced scientific world in which we all live.

What we offer, then, is an effort to share some small portion of our learning — a portion that we hope will be of interest and benefit to others. In this respect it is worth noting that humanitarian engineering education can have benefits well beyond immediate contexts of humanitarian crisis and need. In a progressively globalized world, the successful pursuit of industrial activity and corporate enterprise will require increased sensitivity to societal and cultural issues — precisely the kind of sensitivity that should be inherent to any humanitarian engineering teaching and learning. In the field of government service as well, the development of skills associated with humanitarian engineering can be particularly beneficial. For those students who seek to practice humanitarian engineering directly, it can be projected that numerous non-governmental organizations or NGOs will increasingly depend on the abilities of students who have contributed to and graduated from such programs.

In her contribution to a special issue of the *Journal of Engineering Education*, Shirley Malcolm (2008), Head of the Directorate for Education and Human Resources at the American Association for the Advancement of Science, presented what we call humanitarian engineering as one example of the human face of engineering. But at the risk of over reaching, we would dare to suggest more: that humanitarian engineering may also be the human face of humanity.

Figure 6.1: Emilie Horne, "Without Being Grandiose" (2008), acrylic and pencil on panel 12" × 12", an aesthetic expression of the humanitarian engineering ideal.

In a scholarly reflection on the historico-philosophical career of humanity in the 20th century, the British philosopher Jonathan Glover reviewed the decline of the traditional institutions of morality, especially religion and belief in the transcendent; examined what we know about the moral psychology of warfare and how the great wars of the last century came about; noted the persistence and evil consequences of tribalism; summarized how terror functioned to debase human beings under Stalin and Mao, and how the Nazi experiment to create a master race led to holocaust and ruin. His goal, he wrote, was "an attempt to give ethics an empirical dimensions [by using] ethics to pose questions to history and ... history to give a picture of the parts of human potentiality which are relevant to ethics" (Glover, J., 1999, p. x). Accepting the absence of what he terms external moral law, he argues that "morality needs to be humanized: to be rooted in human needs and human values" (Glover, J., 1999, p. 406). Conspicuous by its absence in his narrative, however, is any discussion of engineering or of humanitarianism — or even of Abraham Maslow and hedonic psychology.

One of the most sobering experiences of the previous century has been the persistent tendency of technology to be put to destructive or superficial uses, both of which — although in ways not to be equated — have diminished our humanity. Although we noted in Chapter 2 that there is a distinction between humanism and humanitarianism, we conclude by venturing to suggest that humanitarianism and humanitarian engineering constitute a positive, no matter how partial response to the problem Glover poses.

Bibliography

This list of references does not include classic authors or texts for which standard citation is sufficient to clearly identify a source. Neither does it include basic reference works that are otherwise adequately identified in the main body of the text.

Ajango, Deb, ed. (2005) *Lessons Learned II: Using Case Studies and History to Improve Safety Education.* Palm Springs, CA: Watchmaker Publishing. 53

Albertson, Maurice L., Andrew E. Rice, and Pauline E. Birky. (1961) *New Frontiers for American Youth: Perspectives on the Peace Corps.* Washington, DC: Public Affairs Press. 33, 41

Albritton, Jane. (2008) Study Water, Save the World with a Volunteer Corps, *Northern Colorado Business Report* (August 18). 32

Anderson, Scott. (1999) *The Man Who Tried to Save the World: The Dangerous Life and Mysterious Disappearance of Fred Cuny.* New York: Doubleday, 1999. 28

Architecture for Humanity [Kate Stohr and Cameron Sinclair, eds.]. (2006) *Design Like You Give a Damn: Architectural Responses to Humanitarian Crises.* New York: Metropolis Books.

Arenson, Karen W. (1995) "Missing Relief Expert Gets MacArthur Grant," *New York Times*, June 13, 1995, p. A12.

Azapagic, Adisa, Slobodan Perdan, and Roland Clift, eds. (2004) *Sustainable Development in Practice: Case Studies for Engineers and Scientists.* Chichester, UK: John Wiley. 50

Baillie, Caroline. (2006) *Engineering within a Local and Global Society.* San Rafael, CA: Morgan and Claypool. DOI: 10.2200/S00059ED1V01Y200609ETS002 44

Baillie, Caroline, and George Catalano. (2009) *Engineering and Society: Working Towards Social Justice.* 3 vols. San Rafael, CA: Morgan and Claypool. DOI: 10.2200/S00136ED1V01Y200905ETS008

Barnett, Michael, and Thomas G. Weiss, eds. (2008) *Humanitarianism in Question: Politics, Power, and Ethics.* Ithaca, NY: Cornell University Press. 51

Biagioli, Mario, ed. (1999) *The Science Studies Reader.* New York: Routledge. 1

Biggs, Stephen. (1989) *Resource-poor Farmer Participation in Research: A Synthesis of Experiences from Nine National Agricultural Research Systems.* OFCOR Comparative Study Paper 3. The Hague: International Service for National Agricultural Research.

Bridger, J. C., and A. E. Luloff. (1999) "Toward an Interactional Approach to Sustainable Community Development," *Journal of Rural Studies*, vol. 15, no. 4 (October), pp. 377–387. 44

Cahill, Kevin M., ed. (2005) *Technology for Humanitarian Action.* New York Fordham University Press and the Center for International Health and Cooperation. 34

Catalano, George. (2007) *Engineering, Poverty, and the Earth.* San Rafael, CA: Morgan and Claypool. DOI: 10.2200/S00088ED1V01Y200704ETS004

Cohen, Lou. (1995) *Quality Function Deployment: How to Make QFD Work for You.* Englewood Cliffs, NJ: Prentice Hall. 42

Cuny, Frederick C. (1983) *Disasters and Development.* New York: Oxford University Press. 27, 29, 30

Cuny, Frederick C., with Richard B. Hill. (1999) *Famine, Conflict and Response: A Basic Guide.* West Hartford, CT: Kumarian Press. 30

Davis, Jan, and Robert Lambert. (2002) *Engineering in Emergencies: A Practical Guide for Relief Workers*, 2nd edition, Warwickshire, UK: Practical Action Publishing. 43

Davis, Michael. (1998) *Thinking Like an Engineering: Studies in the Ethics of a Profession.* New York: Oxford University Press, 1998. 1

Dean, Kari Lynn. (2004) "A MacGyver for the Third World," *Wired*, 10.11.04. 42

DiPrizio, Robert C. (2002) *Armed Humanitarians: U.S. Interventions from Northern Iraq to Kosovo.* Baltimore: Johns Hopkins University Press. 24

Fals-Borda, Orlando and Mohammad Rahman, eds. (1991) *Action and Knowledge: Breaking the Monopoly with Participatory Action Research.* New York: Apex Press. 44

Forsythe, David P. (2005) *The Humanitarians: The International Committee of the Red Cross.* Cambridge, UK: Cambridge University Press. 20

Freire, Paolo. (1970) *Pedagogy of the Oppressed.* New York: Herder and Herder. 44

Galtung, Johan, and Katrin Lederer. (1980) "Areas for Further Research in Needs." In Katrin Lederer, ed., *Human Needs: A Contribution to the Current Debate* (Cambridge, MA: Oelgeschlager, Gunn, and Hain), pp. 345–346. 48

Glover, Jonathan. (1999) *Humanity: A Moral History of the Twentieh Century.* New Haven: Yale University Press. 59, 62

Goldman, Steven L. (1991) "The Social Captivity of Engineering." In Paul T. Durbin, ed., *Critical Perspectives on Nonacademic Science and Engineering* (Bethlehem, PA: Lehigh University Press), pp. 121-145. 35

Habermas, Jürgen. (1968) *Erkenntnis und Interesse*. Frankfurt am Main: Suhrkamp. English versions: *Knowledge and Human Interests*. Trans. Jeremy J. Shapiro. Boston: Beacon Press, 1971. 48

Hofstede, Geert. (1984) "The Cultural Relativity of the Quality of Life Concept," *Academy of Management Review*, vol. 9, no. 3, pp. 384–398. 48

Illich, Ivan. (1967) "The Seamy Side of Charity," *America*, vol. 116, no. 3 (January 21), pp. 88–91. 57

Illich, Ivan. (1978) *Toward a History of Needs*. New York: Pantheon. 49

Institution of Civil Engineers. (2008) "Code of Professional Conduct." London: ICE. 5, 7

International Union for Conservation of Nature and Natural Resources. (1980) *World Conservation Strategy: Living Resource Conservation for Sustainable Development*. Gland, Switzerland: IUCN. 39

Johnston, Stephen, Alison Lee, and Helen McGregor. (1996) "Engineering as Captive Discourse," *Society for Philosophy and Technology Electronic Journal*, vol. 1, nos. 3–4. 35

Kandachar, Prabhu, and Minna Halme, eds. (2008) *Sustainability Challenges and Solutions at the Base of the Pyramid: Business, Technology and the Poor*. Sheffield, UK: Greenleaf. 51

Kahneman, Daniel, Ed Diener, and Norbert Schwarz, (1999) *Well-Being: The Foundations of Hedonic Psychology*. New York: Russell Sage Foundation. 47

Kennedy, David. (2004) *The Dark Sides of Virtue: Reassessing International Humanitarianism*. Princeton, NJ: Princeton University Press. 24

Kennedy, Pagan. (2003) "Necessity Is the Mother of Invention," *New York Times Magazine* (November 30). 42

Koen, Billy Vaughn. (2003) *Discussion of the Method: Conducting the Engineer's Approach to Problem Solving*. New York: Oxford University Press. 2

Malcolm, Shirley M. (2008) "The Human Face of Engineering," *Journal of Engineering Education*, vol. 97, no. 3 (July), pp. 237–238. 61

Maslow, Abraham. (1954) *Motivation and Personality*. New York: Harper and Row. 47

Meadows, Donella H., Dennis L. Meadows, Jørgen Randers, and William W. Behrens III. (1972) *The Limits to Growth*. New York: Universe Books. 38

Minear, Larry, ed. (2002) *The Humanitarian Enterprise: Dilemmas and Discoveries*. Bloomfield, CT: Kumarian Press. 24

Mitcham, Carl. (1994) "Engineering Design Research and Social Responsibility." In K.S. Shrader-Frechette, *Research Ethics*, (Totowa, NJ: Rowman & Littlefield, 1994). pp. 153–168. 7

Mitcham, Carl. (1995) "The Concept of Sustainable Development: Its Origins and Ambivalence," *Technology in Society*, vol. 17, no. 3, pp. 311–326. DOI: 10.1016/0160-791X(95)00008-F

Mitcham, Carl. (2003) "Professional Idealism among Scientists and Engineers: A Neglected Tradition in STS Studies," *Technology in Society*, vol. 25, no. 2, pp. 249–262. DOI: 10.1016/S0160-791X(03)00019-8 27

Mulder, Karel, (2006) *Sustainable Development for Engineers: A Handbook and Resource Guide.* Sheffield, UK: Greenleaf. 50

Peabody, Sue. (2004) "Slavery and the Slave Trade." In Jonathan Dewald, ed., *Europe, 1450 to 1789: Encyclopedia of the Early Modern World* (New York: Scribners), vol. 5, pp. 429–438. 16

Pfatteicher, Sarah K. A. (2003) "Depending on Character: ASCE Shapes Its First Code of Ethics," *Journal of Professional Issues in Engineering Education and Practice*, vol. 129, no. 1 (January), pp. 21–31. DOI: 10.1061/(ASCE)1052-3928(2003)129:1(21) 6

Pörksen, Uwe. (1995) *Plastic Words: The Tyranny of a Modular Language.* Trans. Jutta Mason and David Cayley. University Park, PA: Pennsylvania State University Press. 39

Press Release. (2006) "Maurice L. Albertson, a Founder of the Peace Corps, to Receive Honorary Degree at Colorado State May 12." Department of Public Relations, Colorado State University, May 10. 33

Pritchard, Michael. (1998) "Professional Responsibility: Focusing on the Exemplary," *Science and Engineering Ethics*, vol. 4, no. 2, pp. 215–233. DOI: 10.1007/s11948-998-0052-8 28

Public Broadcasting System. (1997) *Frontline: The Lost American.* 28

Rieff, David. (2002) *A Bed for the Night: Humanitarianism in Crisis.* New York: Simon and Schuster. 23

Riley, Donna. (2008) *Engineering and Social Justice.* San Rafael, CA: Morgan and Claypool. xii

Rostow, Walt W. (1960) *The Stages of Economic Growth: A Non-communist Manifesto.* Cambridge, UK: Cambridge University Press. 9

Sachs, Jeffrey. (2005) *The End of Poverty: Economic Possibilities for Our Time.* New York: Penguin. 23

Schneider, Jen, Juan Lucena, and Jon A. Leydens. (2009) "Engineering to Help: The Value of Critique in Engineering Service," *IEEE Technology and Society Magazine*, vol. 28, no. 4 (Winter), pp. 42–48. 57

Schumacher, E.F. (1973) *Small Is Beautiful: Economics as if People Mattered.* New York: Harper and Row. 30

Sphere Project. (2004) *Humanitarian Charter and Minimum Standards in Disaster Response.* Geneva: Sphere Project. 24, 25, 26

Smith, Preserved. (1957) *A History of Modern Culture.* Gloucester, MA: Peter Smith. 16

Spicer, Edward H., ed. (1952) *Human Problems in Technological Change.* New York: John Wiley. 41

Tanguy, Joelle (1999). "The *Médecins Sans Frontières* Experience." In Kevin M. Cahill, ed., *A Framework for Survival: Health, Human Rights, and Humanitarian Assistance,* (New York: Routledge), pp. 226–244. 22

Terry, Fiona. (2002) *Condemned to Repeat? The Paradox of Humanitarian Action.* Ithaca, NY: Cornell University Press. 24

Tredgold, Thomas. (1828) "Description of a Civil Engineer," *Minutes of the Proceedings of the Institution of Civil Engineers,* vol. 2 (1827–1835), January 4, pp. 20-23. 4

VanderSteen, Jonathan Daniel James. (2008) *Humanitarian Engineering in the Engineering Curriculum.* PhD thesis, Civil Engineering, Kingston, Ontario, Canada: Queen's University. 27, 49, 55

Ward, Lester F. (1883) *Dynamic Sociology,* vol. 2. New York: Appleton. 15

World Bank. (2008) *Global Economic Prospects 2008: Technology Diffusion in the Developing World.* Washington, DC: International Bank for Reconstruction and Development/World Bank. 49

World Commission on Environment and Development. (1987) *Our Common Future.* New York: Oxford University Press. 38

Supplemental Bibliography

The following bibliography, often with brief annotations, points toward other works on which we have drawn more generally than those explicitly referenced, along with an indication of some supplemental resources for further exploration and educational purposes.

Barry, Christian, and Thomas W. Pogge, eds. (2005) *Global Institutions and Responsibilities: Achieving Global Justice*. Malden, MA: Blackwell.

Seventeen papers on global social justice.

Bernard, H. Russell, and Pertti Pelto, eds. (1987) *Technology and Social Change*. 2nd ed. Prospect Heights, IL: Waveland Press.

A set of 13 anthropological case studies of the problems; complements Spicer (1952), cited in the references. For a monograph on this topic, see Pertti J. Pelto, *The Snowmobile Revolution: Technology and Social Change in the Arctic* (Menlo Park, CA: Cummings, 1973).

Bergman, Carol, ed. (2003) *Another Day in Paradise: International Humanitarian Workers Tell Their Stories*. Maryknoll, NY: Orbis Books.

Fifteen case study stories by front line humanitarian workers.

Bilger, Burkhard. "Hearth Surgery: The Quest for a Stove That Can Save the World." (2009) *New Yorker*, December 21 and 28, pp. 84–97.

Black, Richard. (2003) "Ethical Codes in Humanitarian Emergencies: From Practice to Research?," *Disasters*, vol. 27, no. 2, pp. 95–108.

Burkey, Stan. (1993) *People First: A Guide to Self-Reliant, Participatory Rural Development*. London: Zed Books.

An extended case for self-reliant participatory development and participatory action research.

Cahill, Kevin M., ed. (1999) *A Framework for Survival: Health, Human Rights, and Humanitarian Assistance in Conflicts and Disasters*. Rev. ed. New York: Routledge.

Cahill, Kevin M., ed. (2003) Basics of International Humanitarian Missions. New York: Fordham University Press and the Center for International Health and Cooperation.

Ten papers covering historical and practical issues. For more on the same topics, see also Cahill, ed., *Emergency Relief Operations* (New York: Fordham University Press and the Center for International Health and Cooperation, 2003).

Cahill, Kevin M., ed. (2003) *Traditions, Values, and Humanitarian Action.* New York: Fordham University Press and the Center for International Health and Cooperation.

Nineteen historically and philosophically oriented papers.

Cahill, Kevin M., ed. (2004) *Human Security for All: A Tribute to Sergio Vieira de Mello.* New York: Fordham University Press and the Center for International Health and Cooperation.

Sixteen tributes to the life and work to the UN High Commissioner for Human Rights, who was killed by terrorists in Baghdad in 2003.

Escobar, Arturo. (1995) *Encountering Development: The Making and Unmaking of the Third World.* Princeton, NJ: Princeton University Press.

Farmer, Paul. (2005) *Pathologies of Power: Health, Human Rights, and the New War on the Poor.* Berkeley: University of California Press.

Physician, anthropologist, and humanitarian activist Farmer provides eyewitness accounts of his experiences in Haiti, Latin America, and Russia, followed by an extended criticism of neoliberal views of human rights. For an independent profile of Farmer and his work, see Tracy Kidder, *Mountains beyond Mountains* (New York: Random House, 2004).

Hershey, John. (1956) *A Single Pebble.* New York: Random House.

A short, fictional account of a young American engineer scouting out the possibility of a Three Gorges Dam in the 1920s. As he ascends the Yangtze his engineering assumptions about how to help the Chinese are progressively challenged.

Hilhorst, Dorothea. (2005) "Dead Letter or Living Document? Ten Years of the Code of Conduct for Disaster Relief," *Disasters,* vol. 29, no. 4, pp. 351–369.

Hoffmann, Stanley. (1996) *The Ethics and Politics of Humanitarian Intervention.* Notre Dame, IN: University of Notre Dame Press.

Six studies of the problematics of violating national sovereignty in the name of humanitarian action. Includes contributions by Raimo Väyrynen, Robert C. Johansen, and James P. Sterba.

Ishay, Micheline R. (2004) *The History of Human Rights: From Ancient Times to the Globalization Era*. Berkeley: University of California Press.

Sound scholarly overview.

Moon, Suzanne. (2007) *Technology and Ethical Idealism: A History of Development in the Netherlands East Indies*. Leiden, Netherlands: CNWS Publications. Pp. viii, 186.

"Ethical idealists believed that technology could produce development only if it were presented to the indigenous people in the context of a close and trusting relationship" (p. 145). A detailed study of the complexities and often counterproductive results.

Morehead, Caroline. (1999) *Dunant's Dream: War, Switzerland, and the History of the Red Cross*. New York: Caroll and Graf.

A basic history of the International Red Cross/Red Crescent.

National Academy of Engineering. (2004) *The Engineer of 2020: Visions of Engineering in the New Century*. Washington, DC: National Academies Press.

Together with a subsequent study, *Educating the Engineer of 2020: Adapting Engineering Education to the New Century* (2005), argues for "a future where engineers are prepared to adapt to changes in global forces and trends and to ethically assist the world in creating a balance in the standard of living for developing and developed countries alike."

Rist, Gilbert. (1997) *The History of Development: From Western Origins to Global Faith*. London: Zed Books.

A historico-philosophical criticism.

Scott, James C. (1998) *Seeing Like a State: How Certain Schemes to Improve the Human Condition Have Failed*. New Haven: Yale University Press.

A widely referenced study of how massive development schemes fail even in so-called developed countries.

Slim, Hugo. (2000) "Fidelity and Variation: Discerning the Development and Evolution of the Humanitarian Idea," Fletcher Forum of World Affairs, vol. 24, no. 1 (Spring), pp. 5–22.

"Volunteerism and Humanitarian Engineering," part one. *IEEE Technology and Society Magazine*, vol. 28, no. 4 (Winter 2009). Part two, vol. 29, no. 1 (Spring 2010). Two collections of articles, the first guest edited by Kevin M. Passino, the second by Cristelle Didier and Joseph R. Herkert.

Watson, Garth. (1988) *The Civils: Story of the Institution of Civil Engineering*. London: Thomas Telford. Pp. xii, 268.

The most complete story of the British origins of civil engineering.

Weiss, Thomas G. (2007) *Humanitarian Intervention: Ideas in Action.* Malden, MA: Polity Press.

A critical examination of using the military to perform humanitarian relief.

Young, Robert J.C. (2003) *Postcolonialism: A Very Short Introduction.* Oxford: Oxford University Press. Pp. xiv, 178.

Good brief presentation of the attitudes of many intellectuals in the developing worlds of Africa, Asia, and Latin America. Places theory in historical and sociological context.

Authors' Biographies

Carl Mitcham and David Muñoz are colleagues at the Colorado School of Mines as faculty members in the academic divisions of Liberal Arts and International Studies and Engineering, respectively.

CARL MITCHAM

Mitcham earned a PhD in Philosophy from Fordham University, New York. He has taught courses in the ethics of science and technology, while contributing to required core humanities and social science programs at the first and second year levels. Mitcham is editor of the four-volume *Encyclopedia of Science, Technology, and Ethics* (2005) and serves as co-director of an Ethics Across the Curriculum program at CSM.

DAVID MUÑOZ

Muñoz earned a PhD in Mechanical Engineering from Purdue University, Indiana. He has taught courses in thermodynamics, fluid mechanics, and heat transfer, and developed new courses in sustainable engineering. His research interests include energy conservation and issues of engineering design related to global sustainability. Muñoz has done extensive work in Honduras and serves as director of the Humanitarian Engineering minor program at CSM.

CPSIA information can be obtained at www.ICGtesting.com
Printed in the USA
LVOW03s2114100114

368934LV00006B/580/P